Handbook of Hi Fi/Audio Systems and Projects

Also by the Author

Experimenter's Guide to
Solid State Electronics Projects

Practical Guide to Digital Integrated Circuits

Handbook of Hi Fi/Audio Systems and Projects

ALFRED W. BARBER

Parker Publishing Company, Inc. West Nyack, N.Y.

Reward Edition...... September 1983

Library of Congress Cataloging in Publication Data

Barber, Alfred W.
Handbook of hi fi/audio systems and projects.

Includes index.
1. High-fidelity sound systems—Amateurs' manuals. I. Title.
TK9968.B37 621.389'332 81-4536
AACR2

Printed in the United States of America

A Word from the Author on This Book's Practical Value

This book offers two primary benefits: First, it will be invaluable to *all* Hi Fi enthusiasts because it will help you derive greater satisfaction from this rewarding hobby while showing you how to overcome many of the problems encountered in the art of true Hi Fi. Second, this book provides a variety of ingenious, helpful projects you can build to create components that will bring extraordinary improvements to your Hi Fi system. A knowledge and clear understanding of the many techniques described in this book, and your direct involvement, can lead to the highest degree of enjoyment and pride in your Hi Fi system.

Edison thought he had invented Hi Fi 100 years ago. Early radio, talking movies, FM in the 40's and early developers of FM/Stereo all claimed it. I claimed it when I designed the first licensed high fidelity broadcasting station, WQXR. All of these early attempts entertained and even amazed people, but they were limited by the state-of-the-art.

When Lee DeForest invented the vacuum tube, amplification gave high fidelity a strong impulse. For the past 50 years high power, low distortion amplifiers have been available. High quality microphones have been produced for some time. However, electro-mechanical and electro-magnetic components such as phonograph pickups, magnetic recorder/players and loudspeakers have evolved slowly over the years. Perfection still eludes some of the best and most costly efforts. It is now possible, however, to pick up, record and reproduce speech and music to a remarkable degree of faithfulness. The illusion of realism can indeed be attained. How close one can come to the ideal depends on many of the significant factors covered in this book.

To truly enjoy Hi Fi you must first be able to recognize it. Then, appropriate equipment is used to reproduce it. And, finally, a suitable listening environment is needed. If, in addition, you have the satisfaction of having created some of the vital components of your Hi Fi system, you have truly arrived at the height of satisfaction and a much deeper enjoyment. That's what this book is all about.

Chapter 1 introduces you to the world of Hi Fi Projects describing tools and materials most commonly used.Tips on soldering, how to read schematic diagrams and follow wiring diagrams are given. Essential factors affecting Hi Fi, its criteria and its practical limitations are dealt with.

Chapter 2 describes the first Project, a high fidelity 3 watt amplifier including a detailed layout, components used, construction, wiring and testing the finished amplifier. Not only does this building of your first Project result in a remarkable amplifier, but it is also important as a learning phase of Project building, making all subsequent Projects that much simpler to carry out.

Chapter 3 continues with a practical explanation of how much power one needs for really great Hi Fi listening, concluding that 30 watts represents a level at which truly exciting performance can be experienced. This presupposes a reasonably efficient and truly wide response loudspeaker or loudspeakers.

The chapter then continues with details of the design and construction of a 30 watt Hi Fi amplifier and its power supply for mono and stereo use. Experience with this amplifier over a period of time has shown it to be a truly *great* performer.

One can derive musical signals from a record, radio or tape, and amplify these signals to produce a high level of low distortion electrical undulations, but if one cannot adequately convert them to sound waves, all is in vain. The loudspeaker is the all important link between the electrical and the acoustical signals. One must use a good loudspeaker to obtain good sound. There is no other way. (Except, of course, earphones.)

Chapter 4 is devoted to loudspeakers. It describes speaker problems, the significance of the enclosure, how the listening room affects what you hear and many other practical matters. It goes on to describe experiments in developing your own Hi Fi loudspeaker. It describes in detail the construction of a three way, "full size" loudspeaker which will do justice to the best source material you can provide.

Added flexibility for the enjoyment of your Hi Fi system is provided by means of a preamplifier, mixer and equalizer Project described in Chapter 5. Preemphasis in disk and magnetic recording is analyzed. Detailed instructions including circuit diagram, chassis layout and drilling and wiring are given for a compensated preamplifier, suitable for disk or tape, combined with a microphone amplifier and an auxiliary input amplifier *all* of which can be mixed. Helpful ideas are presented for a variety of uses for this preamplifier/mixer. Stereo capability can be provided by building a second identical unit.

Chapter 5 also describes a versatile 5-channel equalizer with full instructions for its wiring and assembly. Again, stereo requires a second equalizer identical to the one described.

You don't need to take someone else's word for performance levels if you can perform your own tests. Chapter 6 describes a number of very useful test Projects you can easily build and use. The most ambitious yet practical Project is an Intermodulation Distortion Meter. With this meter you can measure distortion at levels well below any you can detect by ear in your amplifiers and tape recorders.

Although an oscilloscope can be helpful, a most versatile and interesting Project described is a square wave generator. Its greatest usefulness is probably in evaluating magnetic tapes. This is especially important since manufacturer's descriptions are so often ambiguous.

The fast response sound level meter Project will find

many interesting uses in your experiments, particularly with respect to loudspeakers.

The three tone test oscillator Project is simple yet it produces a very helpful device for evaluating frequency response of such things as tape recorders and loudspeakers.

One of the distinct advantages of building your own equipment is that a clear understanding makes it much easier to trouble shoot. Even the most experienced engineer makes mistakes and must then find the trouble and correct it. Chapter 7 discusses trouble shooting in general and then gives some specific advice regarding your Projects. This chapter also includes such things as measuring the damping factor and the power to your loudspeaker.

Hi Fi record playing depends on many factors. Among these are turntable design, tonearm characteristics, cartridge response and last but not least, the records themselves. Chapter 8 will help you choose the best within your budget. As with most Hi Fi components, the most expensive is *not* necessarily the best nor is it necessary to pay a high price for really good performance.

The most complete source of recorded music is, of course, the perennial phonograph record. In order to get the most out of your records, you need really good reproducing equipment. Chapter 8 will help you judge turntables, tonearms and cartridges. Guidelines are also given on how to select records for your Hi Fi system. Test records are available and are accompanied by detailed instructions for their use.

The most challenging device in the Hi Fi repertory is the magnetic tape recorder. This important development has come a long way and is still the subject of intensive study and interest. Chapter 9 provides practical data on magnetic tape recording in all of its many aspects.

Subjects such as the High Frequency Characteristics of Magnetic Recording, Open-Reel vs. Cassette Recording, Choosing a Cassette Deck and Choosing a Tape for Your Deck are elaborated on. Many experiments are outlined including Testing Tape Surface Quality, Equalization of Magnetic Tape Recording, Evaluating Your Tapes, and Listening Techniques. The chapter continues with notes on Bias

Adjustment, Dolby Noise Reducers and Speed Adjustment. By studying this material and carrying out the experiments you will obtain the best results from your equipment and most importantly learn to appreciate superb recording and sound reproduction.

The real meaning and deepest enjoyment of Hi Fi comes, in the final analysis, from listening to it. Chapter 10 discusses Listening Tests in General, the Listening Room, Experimenting with Speaker Locations and the various effects of walls—reverberation or too much sound absorption in the listening room. The loudspeaker and the listening room are crucial elements in any Hi Fi system; they should be given a great deal of attention and experimentation before you decide that you have the best attainable listening environment.

It has been stated earlier that square wave testing is both interesting and *informative*. Chapter 11 is devoted to the art of square wave testing. The ideal square wave is shown, together with waves illustrating various system responses. There follows a series of actual oscillograph pictures of square wave responses from tape to loudspeaker.

Now, begin your adventure into Hi Fi Projects and experiments. You will derive great enjoyment, pride and satisfaction in the fascinating world of high fidelity, as you begin to apply the practical ideas and suggestions contained in this book.

Alfred W. Barber

Contents

A Word from the Author on This Book's Practical Value 5

1. How to Achieve Greater Hi Fi with the Projects in This Book 17

Setting Up Criteria for Hi Fi 17
Factors Limiting Hi Fi 20
Tools You Need for Project Building 23
Suggestions for Useful Power Tools 23
Materials Used for Projects 24
Successful Soldering Tips and Techniques 24
Reading Schematic Circuit Diagrams 25
How Wiring Is Represented 28
Following the Signal Path 29

2. Building a 3 Watt Hi Fi Amplifier and Power Supply 31

Outlining the Amplifier/Power Supply Project 31
Starting with the Power Supply 32
Breadboarding the Power Supply 32
Testing the Power Supply 34
Choosing an Amplifier Circuit 36
Practical Considerations of the Amplifier Circuit 37
Breadboarding the Low Power Amplifier 42
Testing the Low Power Breadboard Amplifier 44
Enclosing the Low Power Amplifier 45
Adding a Stereo Low Power Amplifier 51

3. How to Build a 30 Watt Hi Fi Power Amplifier ... 53

How the Power Supply Determines Maximum Power 53
How Much Power Is Needed? 54

(con't)

Getting the Most Out of a 25 Watt Amplifier 54
Design a 30 Watt Hi Fi Amplifier 57
Build the Power Supply for the 30 Watt Amplifier 59
Build the 30 Watt Amplifier 61
Enclosing the 30 Watt Amplifier 63

4. **Building a Full Size Hi Fi Loudspeaker 67**

What Is the Ideal Loudspeaker? 67
The Listening Room Is Important 68
The Significance of Speaker Enclosures 68
How to Provide Wide Frequency Coverage 69
Harmonic Distortion in Loudspeakers 69
Selecting Loudspeakers for Experimenting 70
Choosing the Bass-Reflex Enclosure 70
Measure Free Air Characteristics 71
Mount Woofer on a Flat Baffle 74
What Is an Infinite Baffle? 74
How Does a Closed Box Affect Speaker Performance? 76
Build a Full-Size Hi Fi Speaker 76
Constructing the Enclosure 77
Adding Sound-Deadening Material 79
Venting the Enclosure 80
Testing the Woofer in the Enclosure 81
Add Midrange and Tweeter 81
How to Use Cross-Over Networks 81
Design Low-Pass and High-Pass Filters 84
Building Smaller Speakers 90
Understanding the Listening Room 90

5. **How to Build Preamplifier, Mixer and Equalizer Projects 93**

Why Pre-emphasis Is Used in Disk Recording 93
Similarities Between Disk and Magnetic Tape Recording 94
Make a Compensated Preamplifier for Tape and Disk 96
How to Prevent Low Frequency Disturbances 96

(con't)

Assembling the Compensating Preamplifier 99
Checking Overall Performance of Disk and Tape Circuits 100
Mixing Sound Signals from Various Sources 103
Editing a Magnetic Tape 103
Brief Description of the Mike, Phono, Tape Equalizer/Mixer 104
Assembling and Wiring the Mixer Board 106
Testing the Mixer 106
Assembling the Mixer in Metal Cabinets 107
Build a Five Channel Equalizer 108
Putting the Equalizer Together 110

6. Building Test Equipment Projects 117

The Significance of Distortion 117
Measuring Distortion 119
Build an Intermodulation Meter 121
Outlining the Basic IM Meter Circuit 122
Intermodulation Meter Circuit Details 123
Assembly and Wiring of an Intermodulation Distortion Meter 127
Mount the IM Distortion Meter in a Cabinet 129
Calibrating the IM Meter 133
Measuring IM Distortion with the Meter 134
Square Wave Testing Reveals Response Characteristic 137
Build a Square Wave Generator 138
Cathode Ray Oscilloscope—Don't Be Without One 141
Tools for the Cathode Ray Oscilloscope 142
Build a White Noise Generator 143
Choose a Sound Level Meter Design 144
Breadboarding the Sound Level Meter 145
Build a Fast Response Sound Level Meter 147
Three Tone Test Oscillator 151
Choosing the Circuit 151
Building the Three Tone Test Oscillator 152
Applying the Three Frequency Test Oscillator 155
Build an Input/Output AC Meter 155

7. Testing and Trouble-Shooting Your Projects 159

Step-by-Step Trouble-Shooting 160
Test Your 3 Watt Amplifier 161
Measure Damping Factor 163
Measuring the Power to Your Loudspeaker 165
Test Your 30 Watt Amplifier 167
Test Your Loudspeaker 168

8. How to Select and Test Turntables, Tone Arms and Cartridges 171

Characteristics of a Good Record Player 171
What to Look for in a Tone Arm 173
Aids in Selecting Your Cartridge 175
Finding Good Records 177
Using Test Records to Fine Tune Your System 178

9. How to Get the Most Out of Your Tape Deck 181

High Frequency Response of Magnetic Recording 182
Open-Reel vs. Cassette Recording 185
Selecting a Fine Cassette Deck 185
Choosing a Compatible Tape for Your Deck 187
Testing Tape Surface Quality 188
Equalization of Magnetic Tape Recording 191
Evaluating Your Tapes 194
Listening Techniques 194
Bias and Bias Adjustment 196
Dolby Noise Reducers Not All the Same 199
Speed Adjustment Assists Copying 200
Care of Your Deck for Long Life 201
Tape Counters: What Do They Really Mean? 201
Tape Surface Testing with IM Meter 203

10. Key Factors That Contribute to Higher Fidelity ... 205

Listening Tests in General 205
The Listening Room Is Important to Hi Fi 206
Experimenting with Speaker Locations 208
Reverberation in a Room Can Be Too Much or Too Little 209

(con't)

Speaker Placement for True Stereo Effects 210
The Ideal Listening Room—How Close Can You Come to It? 211

11. Understanding Square Wave Test Results 215

Glossary 223

Index 231

Handbook of Hi Fi/Audio Systems and Projects

1

How to Achieve Greater Hi Fi with the Projects in This Book

Setting Up Criteria for Hi Fi

The ultimate objective of Hi Fi is to create an illusion of presence. The story has been told that when WJZ first went on the air, high quality sound was expected because measurements showed that the audio response of the transmitter was essentially flat. It sounded terrible, however, because of harmonic distortion that had not been investigated. My criteria for WQXR, the first broadcasting station licensed for High Fidelity, were a flat frequency response within plus or minus 2 dB from 20 to 20,000 cycles and less than 2 percent distortion up to 100 percent modulation. While not stated in the specifications, I was also concerned with studio acoustics, turntable rumble and other factors. The complications of magnetic recording had not been envisioned.

Now, years later, what have we accomplished and what should we expect? Where should we draw the line in defining Hi Fi? Let us consider the components of a Hi Fi Audio System and, in particular, those components which an experimenter can build as projects. I will include test and measuring equipment since building a fine audio amplifier is an accomplishment, but being able to measure its characteristics brings a great deal of added satisfaction.

Ideally, Hi Fi would mean sound reproduction so real that one could not distinguish between the real sound and the reproduced sound. This is a subjective test made after all the tests for frequency response, distortion, etc. have been made and accepted. It evaluates factors not readily measured by instruments. Unfortunately some of these factors cannot be controlled at the reproducing end of the system, such as the feeling of "presence" of a symphony orchestra in a big hall or the call of a bird out-of-doors on a tree top. One is limited by the acoustics of one's listening room.

I suggest, as a starting point, the following characteristics:

Audio Amplifiers:

Frequency Response: Flat to within ± 2dB from 20 Hz to 20 kHz.
Distortion: Less than 2 percent for any type
Power output: 20 watts per channel, however there is no magic number here
Signal-to-Noise Ratio: 90dB minimum
Damping Factor: 20 minimum

Radio Tuners:

AM: Not possible with present 10kHz channel assignments. (WQXR had a 20kHz channel assignment for High Fidelity)
FM:
Quieting: 15dBf/30dBf
Signal-to-Noise Ratio: 75dB minimum
Distortion: less than 1 percent
Separation: greater than 40dB at 1kHz
Capture: less than 2dB
Selectivity: greater than 70dB

Turntables:

Rumble: not more than –70dB (DIN B)
Wow/Flutter: Less than 0.06 percent (weighted per WRMS std.)
Speed: Synchronous or ± 6 percent adjustable

Tone Arms and Cartridges:

VTF (Vertical Tracking Force): 0 to 3 grams
Antiskating: 0 to 3 grams
Tracking Error: less than 1.5 degrees
Cuing: Desirable
Response: 20Hz to 20kHz ± 2dB
Separation: Greater than 30dB at 1kHz
Vertical Angle: 15-20 degrees

Loudspeakers:

Note: It is not possible to define High Fidelity speakers in terms of a few characteristics since there are so many types and variations. The following should be considered as a simple set of guidelines for starters:
Type: Three-way electrodynamic
Response: 20Hz to 20kHz ± 2dB
Impedance: Nominal 8 ohms
Maximum power: 100 watts

Open-Reel Tape Machines:

Speeds: 1⅞ to 15 inches per second
Record/Playback Response: 20Hz to 20kHz ± 2dB
Flutter: Not more than 0.1 percent (WRMS) at 3¾ ips
Signal-to-Noise Ratio: at least 60dB
Distortion: less than 2 percent at 0 VU recording level
Meters: two VU plus LED peak indicators

Cassette:

Same as for Open-Reel above except speeds.

Note on tape recorder/players: High Fidelity performance is only possible with premium tapes. Low grade tapes limit frequency response, dynamic range and signal-to-noise ratio in tape machines. Also the bias and equalization used in recording must match the tape being used.

Factors Limiting High Fidelity

Audio amplifiers are about the only component in an overall high fidelity system which can be built to and beyond any reasonable criteria for high fidelity. Also important is the fact that practically all of their performance characteristics can be measured and specified. However, due to limitations imposed by other components, one does not need the ultimate in amplifier performance. For example, if one has a loudspeaker rated at a maximum input of 25 watts, the use of a 200 watt amplifier is unnecessarily expensive and is actually hazardous.

Radio reception is mainly determined by the transmitting station. Radio receivers are readily available which take almost full advantage of what they receive. AM radio transmitters are limited by the width of their assigned channels. Standard AM broadcast channels in the United States are 10KC wide. The side-bands are thus limited to 5KC on each side of the carrier. To attempt more would cause the side-bands to impinge on the next channels on both sides. If reception of these wider side-bands is attempted, interference from the side-bands in the adjacent channels may occur. Some stations overmodulate and illegally cause interference in the next channels even in receivers which have very good selectivity in receiving only the desired channel. Thus from a practical standpoint AM reception is limited to signals providing no more than 5KC maximum audio frequencies.

FM, on the other hand, is capable of transmitting much higher frequencies. FM stations in the United States are assigned 200KC channels permitting deviations of ±75KC. These parameters will permit audio modulation up to, say, 20KC, however, with increasing distortion toward the upper limit. Since the harmonics of components above 15KC are inaudible, this distortion is of little consequence. Thus, on the bias of transmittable frequency spectrum, FM has a decided advantage.

Both AM and FM can be transmitted and reproduced with low distortion. FM has another advantage which is, that with a suitable signal level at the receiver, FM quiets any noise to a very low level. From some other angles the comparison is somewhat slanted in the other direction. AM can be received over considerable distances and if the signal is strong enough, it has a satisfactory signal-to-noise characteristic. FM has a very limited service range since it is more or less limited to the line of sight from the transmitter to the receiver. In addition, FM signals are readily reflected by buildings and other obstructions causing out-of-phase interference with the desired signal. Also, atmospheric conditions can cause signals from a distance to interfere quite seriously with good AM reception.

To summarize, even though one has a very fine receiver, one is not assured of ideal reception on either AM or FM on all stations and at all times.

Returning to the hi-fi equipment, it has been stated previously that audio amplifiers and radio receivers having a high degree of high fidelity capabilities are available. The limitations of a hi fi system reside mainly in the electro-mechanical devices. The most significant of these are loudspeakers, microphones, phonograph pickups, including the records, and tape recorders including the tape.

Loudspeakers have been the subject of intensive development over the years. Significant progress has been made. However, loudspeakers are difficult to specify and perhaps even more difficult to judge. For example, listening to a loudspeaker demonstration in a dealer's listening room can be quite different from listening to the same loudspeaker in your listening room due to significant differences in the acoustics of the two rooms. Again, speaker placement can make a big difference and for best results, a lot of experimenting may be necessary. When the system is stereo with two speakers, the problem is even more complex. The type of music, too, can have a significant effect as well as the listener's preferences. Small speakers which are fully enclosed are capable of some rather dramatically good reproduction but you just cannot get the deep bass out of a small speaker no matter how good it seems to sound.

Microphones are an important link in the hi fi chain but are generally less important to the experimenter. Very good and often expensive microphones are available, and broadcast and recording studios are generally equipped with quite satisfactory units. It is not difficult to have microphone response substantially superior to some of the other electromechanical components, such as loudspeakers and phonograph pickups in an overall system.

Phonograph pickups and phonograph records can be considered together. They have to work together and there is no particular point in providing one which far excels the other. Actually they both have their limitations. Improved pickups challenge the record making industry. Improved records challenge the art of designing and making pickups. There is room for improvement in both areas.

There are a number of significant steps to be performed in making a final record. Going from a tape master to a direct-to-disk recording does not guarantee a greatly improved record if the vinyl and final pressing are inferior. The best symphony orchestra does not guarantee a superior record if the placement of microphones, mixing and other steps fail to reach high standards. Pickups may be limited in transient response, tracking, dynamic range, etc. Test records are available, but a substantial complement of other test equipment is necessary for definitive tests. Manufacturer's specifications are often incomplete. As to listening tests, one would have to spend a tremendous amount of time with different types of music to reach a valid conclusion. Turntables can introduce wow and rumble.

There are two types of tape recorders which are widely used, namely, open-reel and cassette. The open-reel is the type used by professionals, and open-reel recorders for home use are available. They are capable of very good performance although not reaching the ideal. Magnetic tape recording has a built-in distortion factor ranging from 1 to 2 percent. Distortion can be minimized by reducing the recording level but at the expense of increased noise ratio. The dynamic range also has a practical limit, especially at the high frequencies. Open-reel recorders are large and heavy and rather complicated to operate. They are also expensive. Cassette tape recorders are small, convenient to operate and less expensive. Intensive development has provided operating characteristics which approach those of the open-reel. The tape used for cassette recording becomes a more critical factor since the recording speed is substantially lower. Tape quality is tied in with cost. No cheap tape is very good, particularly when it comes to recording high frequencies. Cheap tape should be definitely used for nothing but the recording of voice (talking, not singing). Again, the better cassette recorders are also more expensive; the difference is most apparent when recording or playing back music.

Tools You Need for Project Building

The following tools are recommended for Project building.

1. Needlenose pliers (small)
2. Diagonal cutters (small)
3. Screw drivers
   a. Very small thin blade
   b. Medium-size
   c. Large
   d. Phillips head small
   e. Phillips head medium
4. Hardened steel or diamond-tipped scribe
5. Adjustable square/rule
6. Medium-size hammer
7. Allen wrenches, small sizes
8. Wire strippers (buy the best) (be sure they will strip No. 24 plastic insulated wire)
9. Soldering iron, fine tip, 40 watts
10. Files, small round, medium flat
11. Hacksaw
12. Small hand-held magnifying glass

With experience, preferences change, and you will find additional tools will be handy, too.

Suggestions for Useful Power Tools

There are two inexpensive power tools which will save a lot of time and effort. These are:

1. Power hand drill
2. Power hacksaw

Along with these a small bench vise is appropriate.

Materials Used for Projects

The small electronic circuits used in many projects are most readily assembled and wired on perf board. This is perforated phenolic board. A board with 0.1 inch hole spacing is appropriate since integrated circuit (DIP) packages and the sockets for DIP packages have pins spaced 0.1 inch apart and so rest neatly on the board with the pins projecting through for wiring on the under side. No. 24 bare tinned copper wire is useful in interconnecting components on the boards. No. 24 stranded plastic insulated wire in various colors is useful in connecting boards to external circuit parts. The colored insulation helps identify circuit connections. A predetermined color code helps even more. For example, the following code could be used:

black for all ground connections
green for signal input
red for positive bias
white for negative bias
blue for output lines

Solder for greatest convenience is rosin core diameter .050 inch.

Successful Soldering Tips and Techniques

Soldering is not really difficult but should be properly done. The following suggestions should help.

1. For projects in this book, a small (30-40 watt) iron is best.
2. Use a small tip such as a ⅛ inch diameter chisel point.

3. Keep the point well tinned and clean (free from black oxidation products). Shake excess solder off the iron and wipe away dirty accumulations. There is a small sponge designed for this purpose.
4. Always wait for the iron to reach a good operating temperature before attempting to solder. Usually in about two minutes solder will flow freely in contact with the tip of the iron.
5. Heat the wires to be soldered for a couple of seconds, apply the rosin core solder between the wire and the soldering iron tip. The solder should flow freely around the joined wires. Don't keep the iron on the wires more than a couple of seconds, especially if the wires go to a semiconductor. Remove the iron and solder. As the solder solidifies its appearance will change from shiny to a sort of silvery mat surface. Inspect the joint with a magnifying glass to see that the solder has wetted all wires in the joint. If it has not, reheat to cause a reflow of solder and if there is too little solder, apply more.

   Remember that poorly soldered joints are a major cause of trouble in solid state circuits.

   Stranded wire should be tinned before use to insure all wires are in place.

Reading Schematic Circuit Diagrams

In order to be able to follow a schematic circuit diagram, it is necessary to know the convention adopted for symbols and their functions. The following is a list of conventional symbols and a brief statement of their functions:

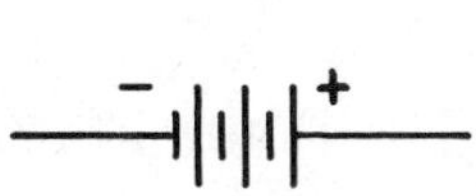

Battery or voltage source; the plus and minus signs indicate polarity

Fixed resistor controls flow of current

Variable resistor

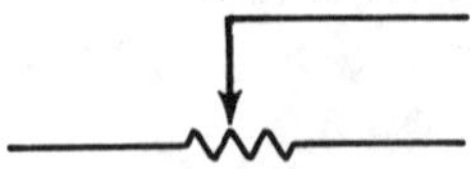

Potentiometer, selects portion of voltage across it

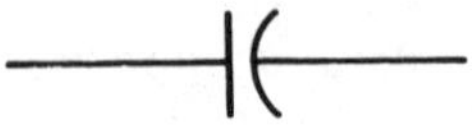

Fixed capacitor, stores electrical energy; breaks a direct current path, filters out ripple in a power supply

Variable capacitor, part of a tuned RF circuit

Inductor, AC impedance, part of a tuned RF circuit

Variable inductor

Iron cored inductor, low frequency impedance

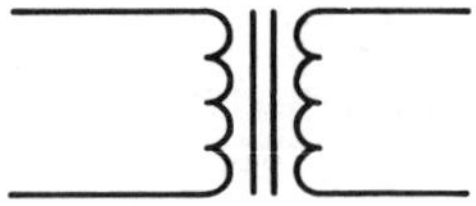

AC transformer, in power circuits changes voltage, in audio circuits changes circuit impedance

Diode rectifier, changes AC to DC in power supplies, radio signal detector

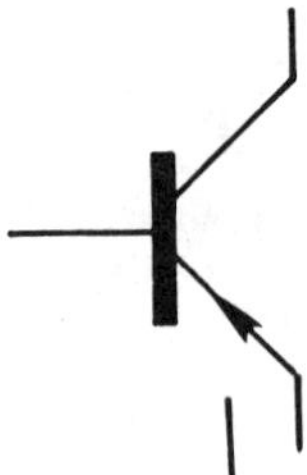

Transistor (PNP) amplifier

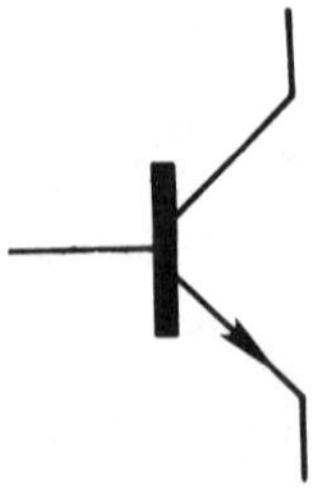

Transistor (NPN) amplifier

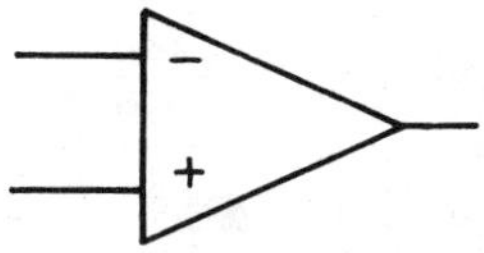

Amplifier such as an operational amplifier

A simple block with an appropriate label is used for devices for which no standard symbol has been applied and for combinations of devices

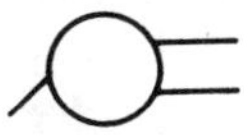

Phonograph cartridge or pickup

Microphone

Alternate ground symbols

Antenna

Two leads joined together

Two leads crossing but not joined conductively

While the above symbols are widely used in this country they are not exclusive. Quite different symbols are used in most foreign countries which makes reading foreign circuit diagrams difficult for the uninitiated.

Almost all of electronics is involved with "circuits." A circuit is a closed loop of circuit elements that carry out some specific function. The simplest circuit is a battery in series with a lamp and a switch (See Figure 1-1). When the switch is open no current flows and there is an "open circuit." When the switch is closed, current flows and there is a "closed circuit."

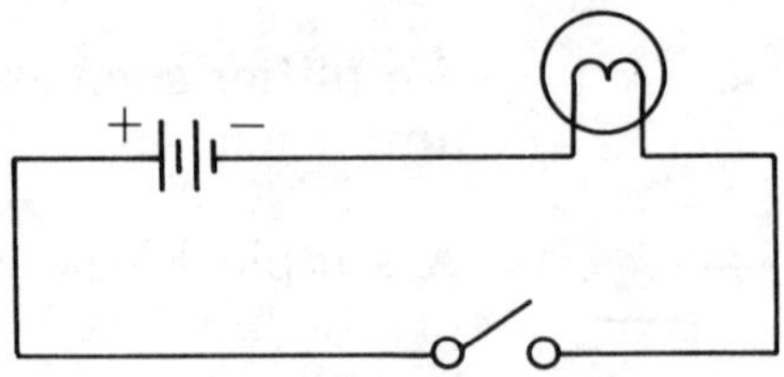

Figure 1-1

In other words, the circuit is a true circuit and performs a function only when the loop is a closed path. Circuits may be simple like the one just described, or more often they may be complex, involving many interconnected loops. A signal to be processed, for example, may be acted upon by a first loop (circuit), passed on to a second loop and acted upon a second time, passed on to a third loop and so on until a final transformation is achieved as when a radio signal is received by an antenna and processed successively until it emerges as sound from a loudspeaker.

The circuits described in presenting the Projects in this book vary from relatively simple circuits to some rather complex circuits. The key to circuit operation is to understand the operation of each step in a chain of functions proceeding from a known source to a desired result. There are actually only a few basic functions which circuits can accomplish. However, when these functions are combined, they can provide an almost infinite variety of overall results. We shall see many of these results in the Projects in this book.

How Wiring is Represented

Wiring diagrams are really simple pictures of a given circuit. They are generally drawn to scale in order to assist in assembly and wiring. Resistors, for example, are shown as small elongated rectangles. Disc capacitors are shown as small circles. Everything is shown to look like the real thing as far as possible in a simple sketch. Component values are often shown next to the actual component. Any panel or chassis drilling is shown to scale indicating the location and size of the holes involved.

Following the Signal Path

The schematic circuit diagram of, for example, a device such as a radio receiver, puts together the appropriate symbols in an orderly, flowing manner. Figure 1-2 is a typical schematic circuit diagram. It is a transistor radio.

By conventional use, all circuit diagrams progress from left to right. In most cases, there is an input at the extreme left. This input signal is processed as it progresses to the right. Finally it emerges at the right end.

In the present circuit, the input is a radio signal received on the antenna. This is a tiny radio frequency signal modulated with an audio signal for speech or music. This signal has no practical use as received and so must be amplified. The transistor Q_1 and Q_2 are radio frequency amplifiers coupled by means of tuned radio frequency transformers. The tiny input signal is thus increased to a substantial level, say one to two volts. This amplified signal is applied to the detector consisting of diode rectifier D. The detector demodulates the amplified signal, in other words, it removes the audio signal from the modulated signal. The R-C filter following the detector removes the radio frequency components leaving the audio, speech or music signal. Transistors Q_3 and Q_4 are audio frequency amplifiers and serve to boost the audio signal to a high level. The amplified audio output of transistor Q_4 is applied to the loudspeaker thus providing the desired speech and music output.

We have now traced the progress of a signal in an amplifier circuit. The tiny radio signal picked up by the antenna is the input. This input is amplified, rectified, amplified again and put out as the desired output signal in the form of music or speech.

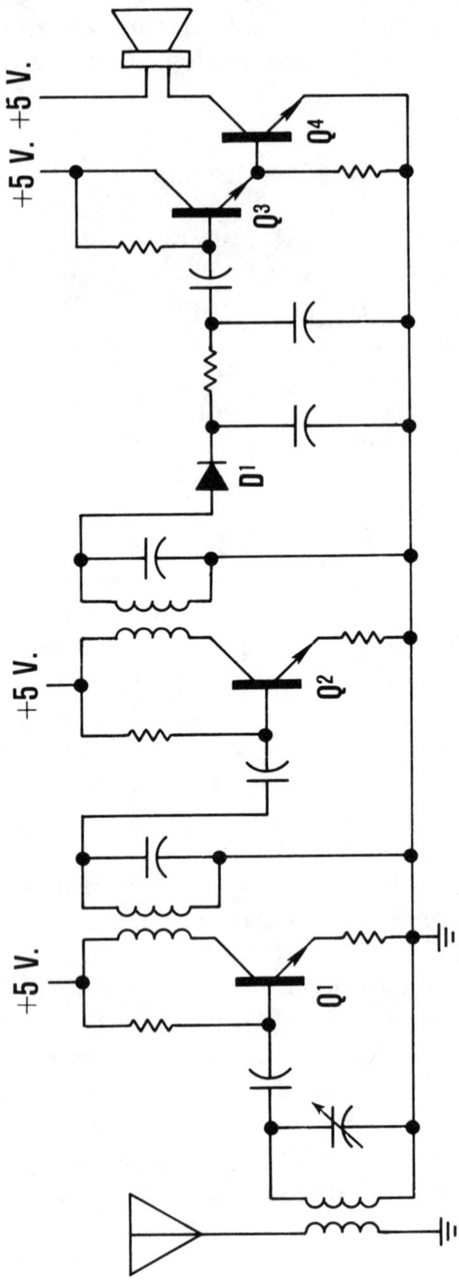

Figure 1-2: Typical Schematic Circuit Diagram

2

Building a 3 Watt Hi Fi Amplifier and Power Supply

A low power amplifier and its power supply is a good starting point for our series of high fidelity projects. This will provide helpful experience in building your own equipment as well as a useful piece of apparatus. The basic wiring, assembly and testing techniques shown here will be used in later projects. The objects of this project are to show how to define objectives, choose the circuits, list the materials and parts required and purchase them, put together a breadboard of the device, test the breadboarded circuit, assemble in a cabinet which gives a finished look, and make final tests.

Outlining the Amplifier/Power Supply Project

The major criteria for an amplifier are: input impedance, gain, power output, frequency response, distortion, load impedance and power requirement. Input and output jacks, on/off switch and gain control should also be included. Finally, the amplifier and its power supply should be mounted in a professional-looking cabinet.

Starting with the Power Supply

A simple power supply built from easily obtainable parts using a standard 120 to 12.6 volts AC at 1.2 amperes stepdown transformer, two 2-3 ampere rectifiers, and four 1000 microfarad 35 volt capacitors in a dual half-wave circuit will provide plus and minus 17.6 volts no load at 120 volts line voltage. Figure 2-1 is a circuit diagram of the proposed power supply.

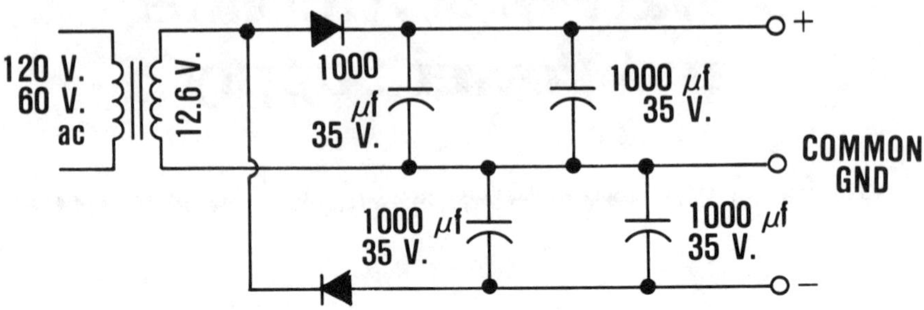

Figure 2-1: Power Supply Circuit Diagram

Breadboarding the Power Supply

The transformer and capacitors were mounted on a piece of perforated board (called perf board) as shown in Figure 2-2. Four 1000 microfarad capacitors with PC (printed circuit) terminals were used. Figure 2-3 shows the assembly and board wiring. Dotted lines indicate under board wires. The following table lists a typical set of components as obtained from Radio Shack although their equivalents are available from other sources. Experimenters and project builders should determine their most available sources of components. In many cases you will find and use more than one source.

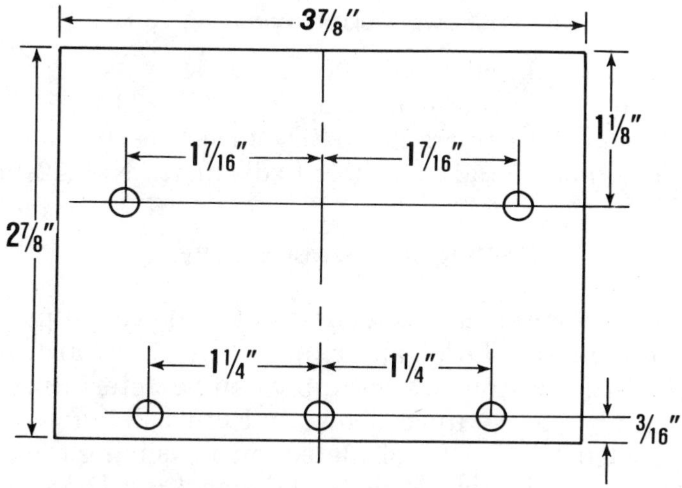

Figure 2-2: Power Supply Perf Board Drilling

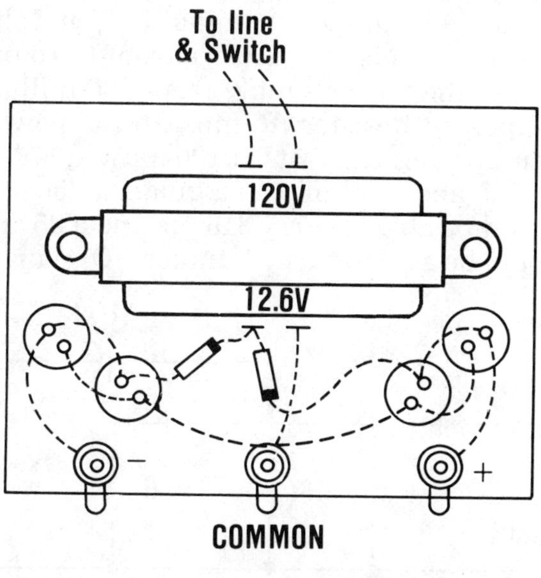

Figure 2-3: Power Supply Wiring Diagram

Power Supply Parts

Perf board Radio Shack No. 276-1394
12.6 volt transformer Radio Shack No. 273-1505
3 ampere rectifiers Radio Shack No. 276-1143
1000 microfarad capacitors Radio Shack No. 272-1032

Testing the Power Supply

The important characteristics of a power supply are output voltage, regulation and ripple. When these are known, the capabilities of the power supply can be determined. The power supply was set up as shown in Figure 2-4 in which the meter labeled E is a DC voltmeter for measuring the power supply output voltage; the meter labeled I is a DC ammeter for measuring the load current; the meter labeled ACVM is an AC voltmeter for measuring ripple; and R is a variable resistance load. Assuming a nominal output voltage of 17 volts, the resistor R must be variable from 170 ohms down to 34 ohms to load the power supply from 100 milliamperes to 500 milliamperes. Resistor R must be a power resistor capable of dissipating 1.7 watts at 100 ma and 170 ohms to 8.5 watts at 500 ma and 34 ohms. As a guide, a 250 ohm 25 watt rheostat should be able to carry 316 ma and a 25 ohm 25 watt rheostat should be able to carry 1 ampere. Of course, data can

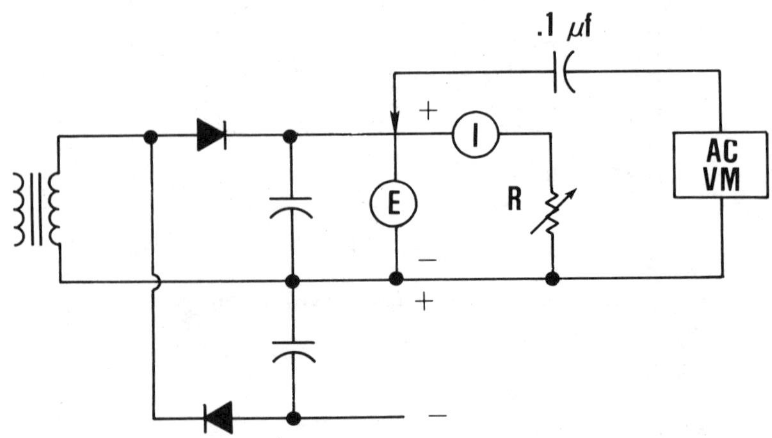

Figure 2-4: Test Set-Up for Testing Power Supply

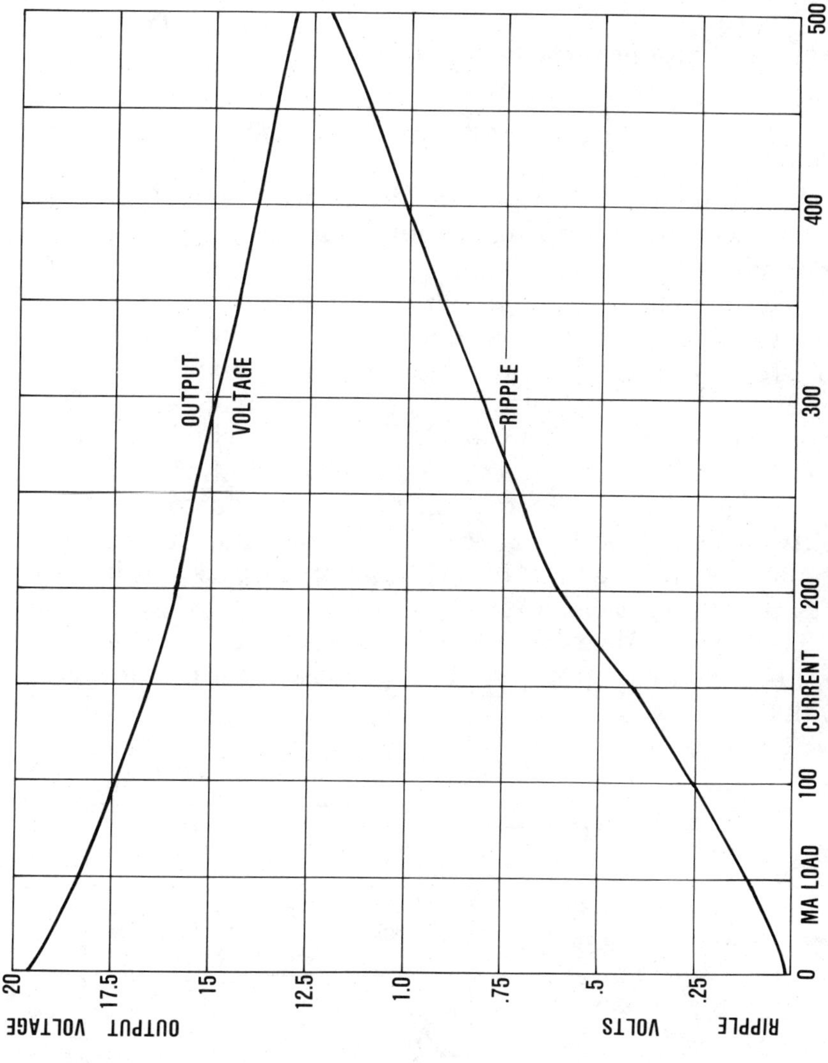

Figure 2-5: Power Supply Regulation and Ripple vs. Load Current

be obtained using combinations of fixed resistors, for example, two 100 ohm 10 watt and one 50 ohm 10 watt resistors can be combined to provide loads of 50, 100, 150, 200 and 250 ohms for taking the data.

Figure 2-5 shows plots of output voltage versus load current and ripple versus load current.

Choosing an Amplifier Circuit

A simple basic amplifier circuit consists of an operational amplifier driving a complementary symmetrical pair of power transistors. Figure 2-6 shows the arrangement. This basic circuit has a number of important features. The speaker load is directly connected to the emitters of the power transistors, eliminating any coupling capacitor and thus providing a flat response down to zero frequency. The feedback is taken from the speaker terminal which provides the best overall frequency and distortion correction as well as the highest damping factor, all of which add to the high fidelity concept. The input to the amplifier is made to the high impedance non-inverting input to the op amp so that it does not load any input circuits.

This circuit does require a double ended power supply but this is no great problem. Our proposed power supply is just that.

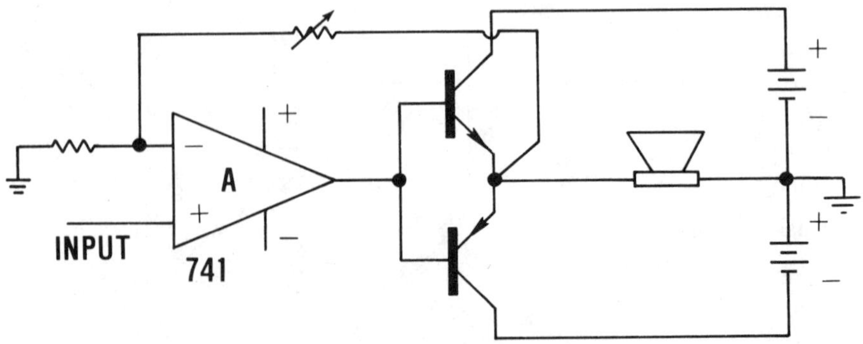

Figure 2-6: Simple Basic Power Amplifier Circuit

Practical Considerations of the Amplifier Circuit

As the proposed circuit is drawn in Figure 2-6, one major problem exists, that is, cross-over distortion. With no input signal the amplifier output is theoretically zero. In a practical case, it is very close to zero provided the amplifier gain is not too high. If the gain is very high, the input off-set voltage and current will bias the output terminal off zero. Figure 2-7 shows how the output off-set varies with closed loop gain, the gain determined by the feedback and input resistors

$$G = \frac{R_{IN} + R_{FB}}{R_{IN}}$$

for four different samples of commercial grade op amps. At gains well below 100, none of these ampifiers showed an output off-set greater than about 0.1 volt which should be tolerable without resorting to zero correction which incidentally, is not always provided for.

In the final design it was decided to provide 3 watts output across an 8 ohm load. This provides approximately 5 volts output and if the input sensitivity is to be 0.5 volt, this means an amplifier gain of about 10 times. As stated above, the amplifier gain in the configuration used is

$$G = \frac{R_{IN} + R_{FB}}{R_{IN}}$$

so that with the 47K feedback resistor and a 4.7K input resistor the gain is 11, which provides 5 volts output for 0.5 volt input and allows for an 0.5 volt drop inserted between the op amp output and the load due to the base to emitter drop in the power transistors. The 25K input potentiometer provides both a gain control and a 25K input impedance to the amplifier. (See Figure 2-10.)

The major problem to be solved in the basic circuit of Figure 2-6 is cross-over distortion. This is also related to the efficiency of the output transistors. For example, if the output transistors are operated Class A, i.e., with a steady collector current equal to half the peak-to-peak collector current, there will be no cross-over distortion but the

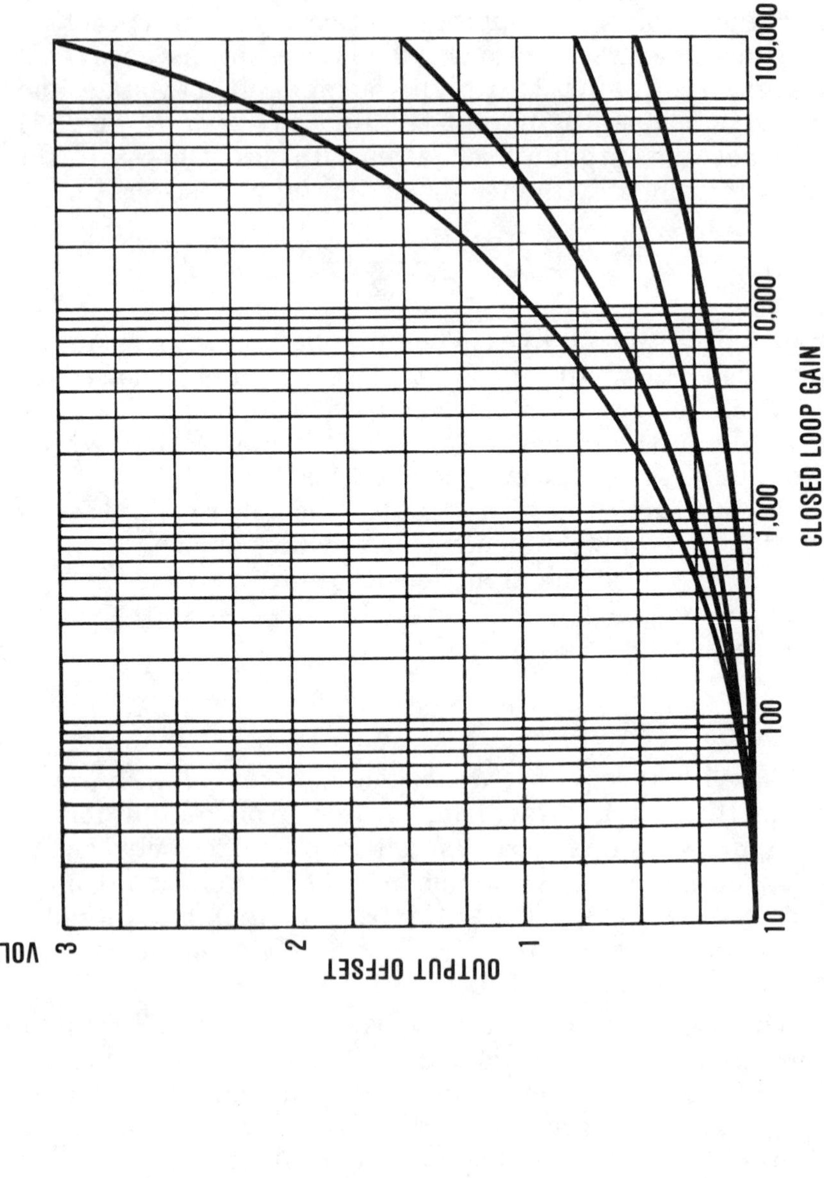

Figure 2-7: Closed Loop Gain op Amp Offset vs. Closed Loop Gain

efficiency will be low. Cross-over distortion takes place where there is no initial bias on the output transistors and the output current dwells at zero until the base voltages in both positive and negative directions exceeds the respective base conduction voltages. See Figure 2-8 where the dotted lines show how the output current does not follow the input voltage around zero. Further explanation is provided in Figure 2-9 which shows that a transistor collector current does not flow at zero base bias but only after a substantial bias is applied (positive for NPN and negative for PNP).

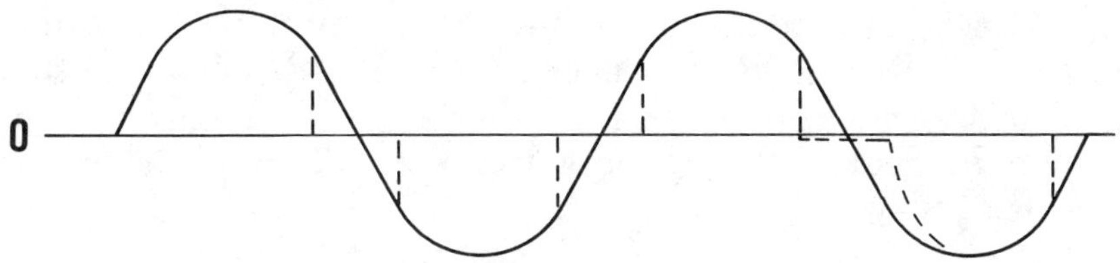

Figure 2-8

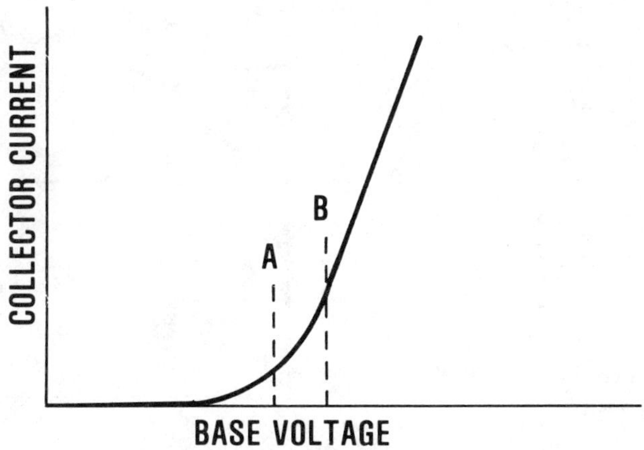

Figure 2-9: Collector Conduction vs. Base Voltage

The conclusion gained from all of this is that the output transistors should be provided with a small initial bias, enough to cause a small current to flow to eliminate cross-over distortion but not enough to adversely affect efficiency. This is not as simple as it might seem since this bias is rather critical and with an unregulated power supply cannot be too accurately determined. One positive aspect, however, is that if the initial bias is on the low side, the feedback will act to reduce the cross-over distortion which might result. The advantage of keeping the initial bias on the low side is that efficiency is improved, heating is reduced, as is any tendency for the power transistors to run away. Figure 2-9 is a plot of transistor collector current versus base bias. It is obvious that a bias at line A will provide a low initial current while a slightly higher bias at line B will provide a substantially higher initial current.

In Figure 2-10 the initial bias is determined by the values of resistor R_1, and R_2 and is equal to $ER_2/(R_1 + R_2)$. Another method of providing initial bias is to substitute diodes for R_2 as shown in Figure 2-11. These two methods operate in two different ways.

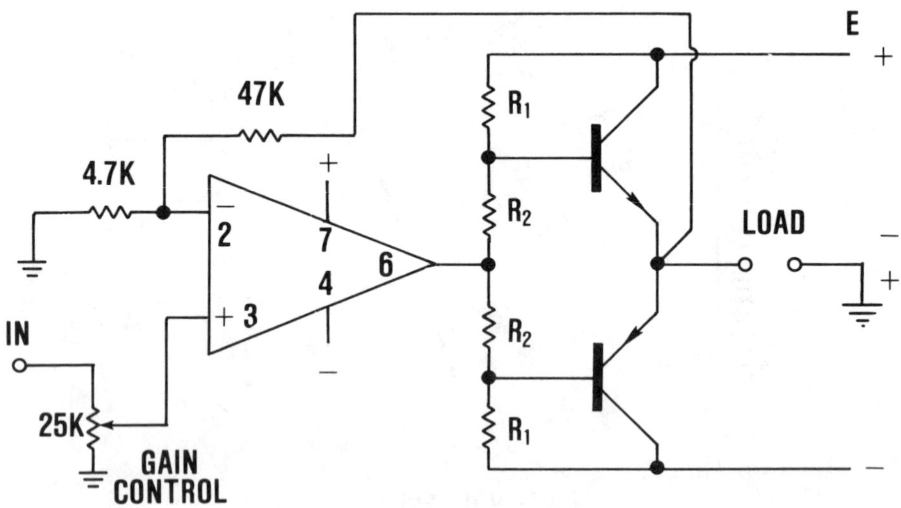

Figure 2-10: Power Amplifier with Bias to Eliminate Crossover Distortion

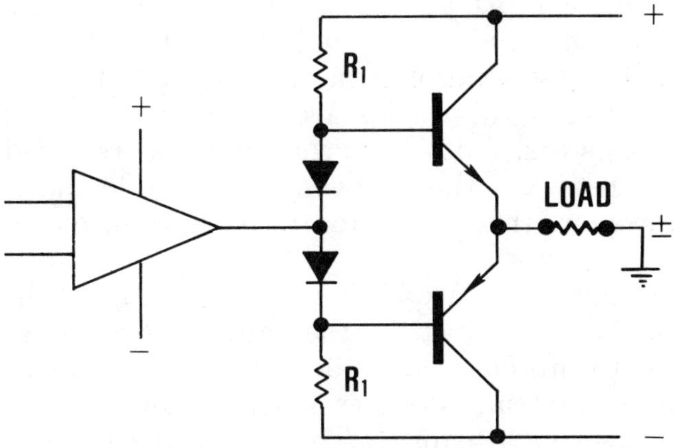

Figure 2-11: Power Transistor with Diode Offset

First, let us examine the diode bias method. With no input signal the diodes are forward biassed, holding the respective output transistor bases at the forward conduction voltages of the diodes, in the case of silicon at about 0.6 volt when a small current is supplied through R_1. However, at 0.6 volt the power transistors are also in conduction, say, at about point B in Figure 2-9. Now when a signal is applied, take for example a positive output going from the op amp, the upper diode is back biassed and current flows to the NPN transistor base through R_1. The output or load current is determined by this base current and hence the maximum power output is a function of R_1. Decreasing the value of R_1 will increase the maximum output but at the expense of higher initial (no signal) current. Experimentally, I observed the following:

R_1	Initial Current	Final Current	Max Output
10K	70 ma	110 ma	0.32 watt
4.7K	150 ma	200 ma	0.78 watt
1K	670 ma	500 ma	3.1 watts

There was no compromise that looked viable. Using power transistors with higher gains would help, but I do not consider this to be the best solution. One advantage of the diode bias method is that if the diodes are thermally coupled to the power transistors, initial current will be reduced if the transistors heat up, thus preventing possible thermal run away. However, this is not too important in a low power amplifier.

Returning to Figure 2-10, the initial bias on the power transistors is determined by the ratio of R_2 to R_1 as stated previously. In this case the value of R_1 is not very significant since the power transistor bases are driven by the op amp through the series resistor R_2. The smaller R_2, the greater the output. However, if R_2 is too small, there will not be enough initial bias to eliminate cross-over distortion. The table below was made from experiments using the R_1 and R_2 method of bias.

R_1	R_2	Initial Current	Final Current	Max Output	Distortion
4.7K	50 ohms	0 ma	360 ma	8 watts	Yes
4.7K	100 ohms	10 ma	360 ma	8 watts	Yes
4.7K	150 ohms	20 ma	350 ma	5.4 watts	No
4.7K	170 ohms	50 ma	350 ma	5.3 watts	No
4.7K	270 ohms	190 ma	330 ma	4.5 watts	No

Note: With $R_2 = 150$ ohms, the initial bias was 0.45 volt, not readily obtainable with diodes.

Breadboarding the Low Power Amplifier

The breadboard was assembled on a 3 + 3¾ inch piece of perf board having 0.1 inch spaced holes using the circuit of Figure 2-10. A low profile ½ DIP socket was used although the op amp could have been mounted directly on the board. Sockets are a convenience especially in experimental projects, since amplifiers can be readily changed and compared or a defective unit quickly replaced. The board was drilled as shown in the dimensioned drawing of Figure 2-12. The transistors, heat sinks and DIP socket were mounted on top of the board as sketched in Figure 2-13. The resistors and interconnections made on the under side of the board are shown in Figure 2-14. Which leads are most conveniently run under the board will then be apparent.

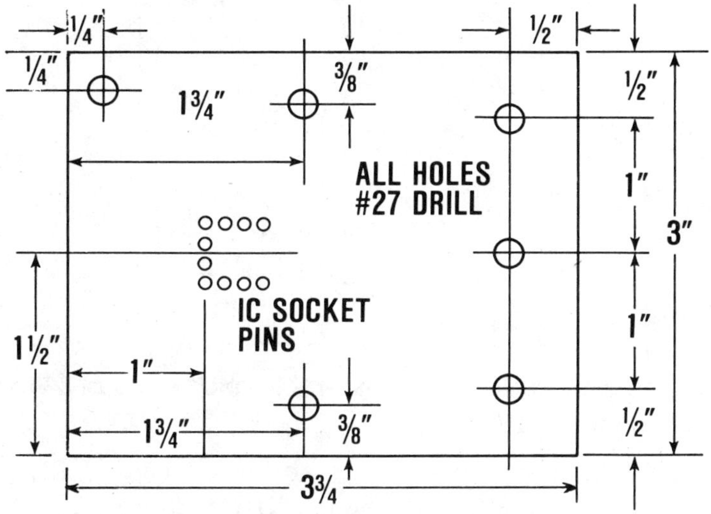

Figure 2-12: 3 Watt Amplifier Perf Board Drilling Layout (Bottom View)

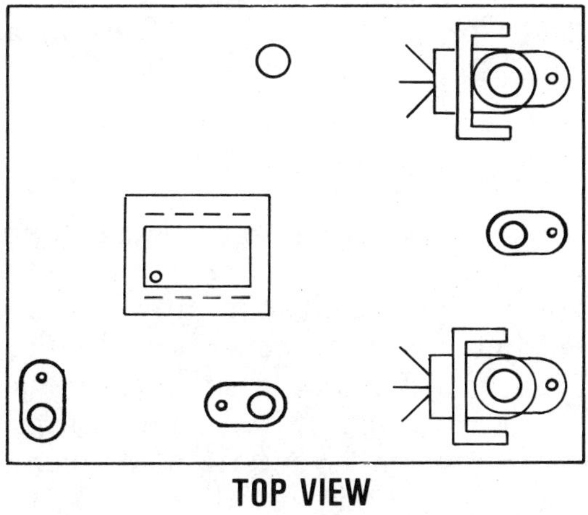

Figure 2-13: 3 Watt Amplifier Layout

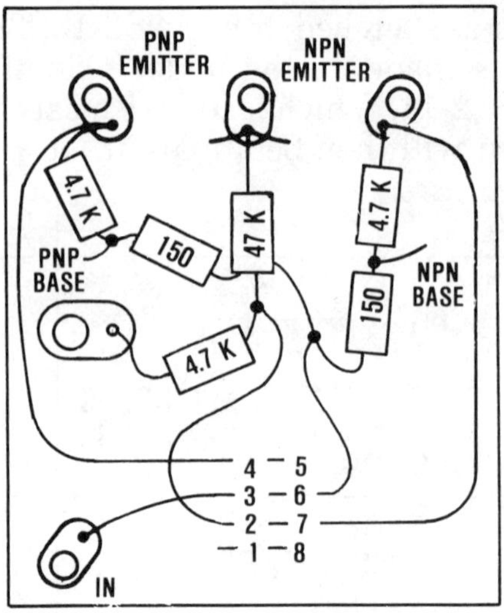

Figure 2-14: Bottom View Assembly and Wiring of 3 Watt Amplifier

The transistors used were the TAB type 2N3719 and 2N3766 available as Radio Shack 276-2025 and 276-2017 or Sylvania ECG Nos. 186 and 187. Radio Shack No. 276-1363 heat sinks were used. The DIP socket was a low profile half DIP. The resistors were all ½ watt carbon (Allan Bradley or equal). They could also be Radio Shack Nos. 271-013, 271-030 and 271-042 (10 percent tolerance). The op amp was an LM1556 (or TL081 Radio Shack No. 276-1716).

Testing the Low Power Breadboard Amplifier

Testing of a high fidelity amplifier includes observing the undue heating of any components particularly the power transistors, checking for any loose connections which could cause noise in the output, and checking the overall performance in general. Specific measurements include sensitivity, maximum undistorted power output, frequency response, harmonic and intermodulation distortion. Which of these measurements an experimenter can make depends on the test equipment that is available.

The present amplifier was measured for frequency response and it was flat within ±1dB from 20Hz to 30kHz. It was down 3.5dB at 70kHz. One word of caution—in making frequency response measurements, be sure to use an AC voltmeter that has a wide flat response or your measurements may be colored by your voltmeter. The measured frequency response is shown in the four-cycle semi-log graph of Figure 2-15. Measured intermodulation distortion is shown in Figure 2-23.

A square wave test will quickly show that the 3 watt amplifier has a much superior response to some other components of any given system, for example, speakers, tape recorders and phonograph pickups. See Chapter 6 for the construction and use of a square wave generator. Figure 2-16 is a cathode ray picture of the response of the amplifier to a 500Hz square wave showing that the output is an essentially exact copy of the input. In later chapters on speakers, tape recorders and phonographs it will become evident that this amplifier is far superior to these latter links in any sound reproduction system.

Enclosing the Low Power Amplifier

With the testing of the breadboard satisfactory and complete the amplifier was mounted in a metal cabinet (Radio Shack Cat. No. 270-0270, size 9¼W × 5⅜H × 6¾D, all in inches). Don't be misled by the fact that the Radio Shack is not consistent in the order in which dimensions are stated nor do they show which dimension is which.

The bottom of the cabinet was drilled as shown in Figure 2-17 to accommodate the power supply and two amplifiers (for stereo). The front panel was drilled as shown in Figure 2-18 and the back panel as shown in Figure 2-19. On the front panel two standard phone jacks, one stereo phone jack, two 25K gain controls, an on/off switch and an LED level indicator were mounted. On the rear panel four phone jacks were mounted. Figure 2-20 is a wiring diagram of all components. Figure 2-21 is a photograph of the inside assembly of the stereo amplifier. Figure 2-22 is a photograph of the finished amplifier.

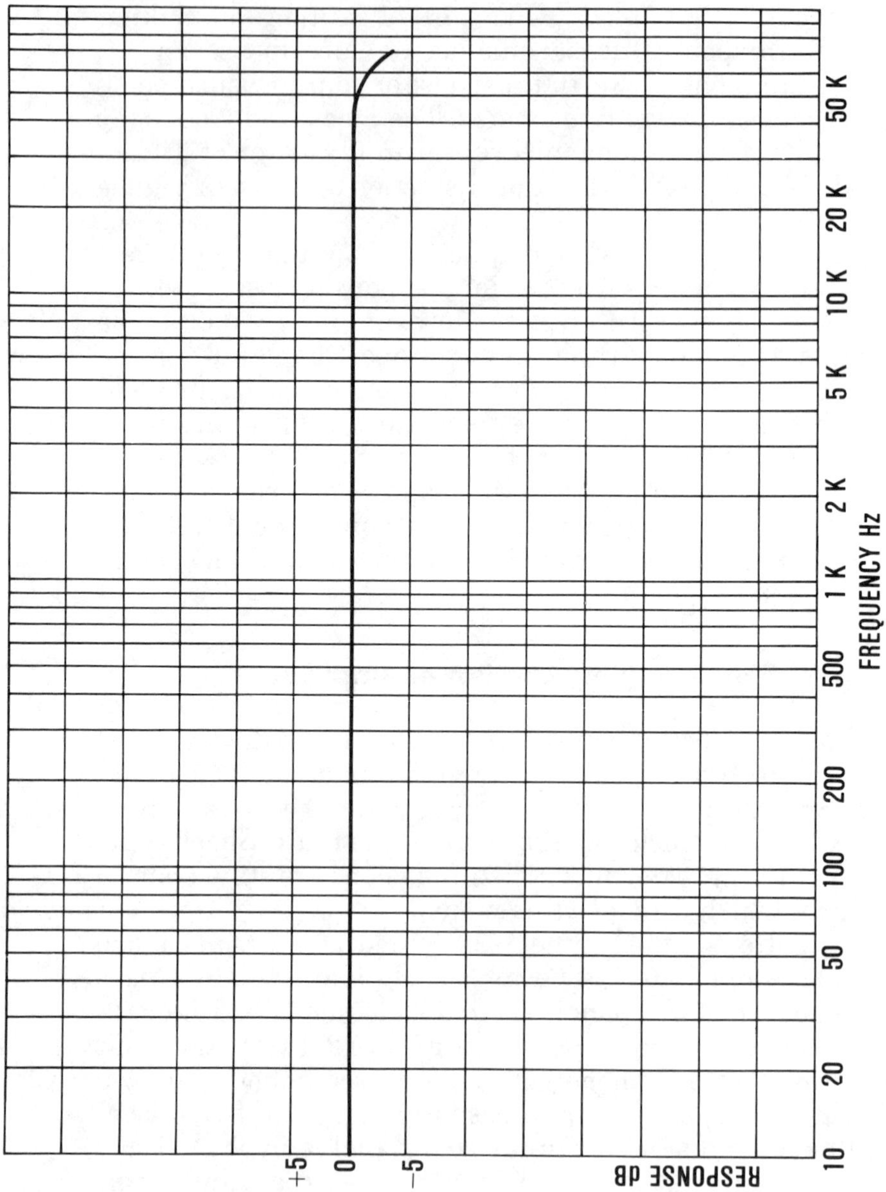

Figure 2-15: Frequency Response of 3 Watt Amplifier

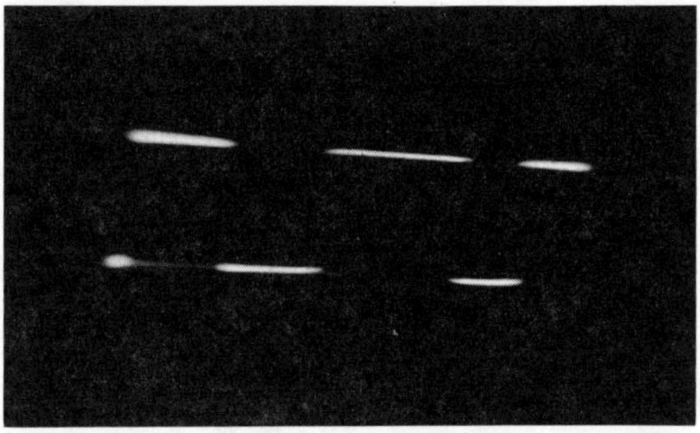

Figure 2-16: Square Wave Response of 3 Watt Amplifier

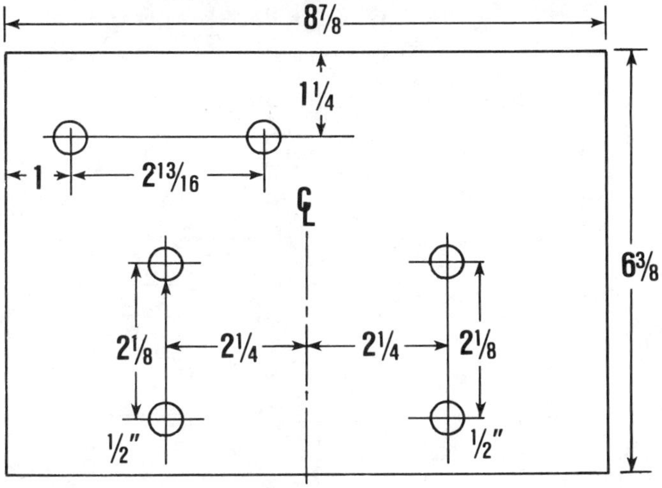

Figure 2-17: Chassis Drilling—Bottom View

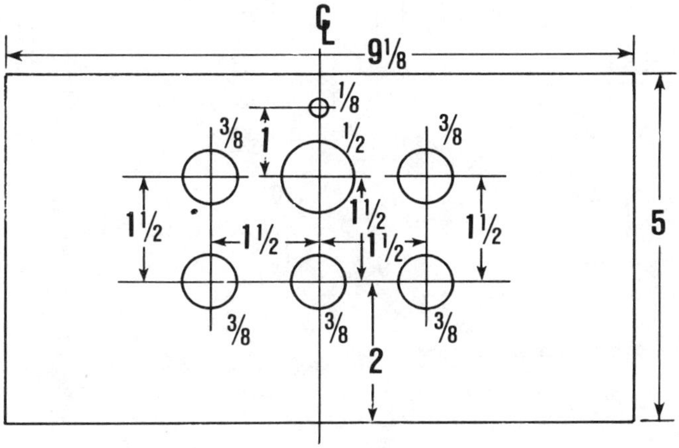

Figure 2-18: Front Panel Drilling

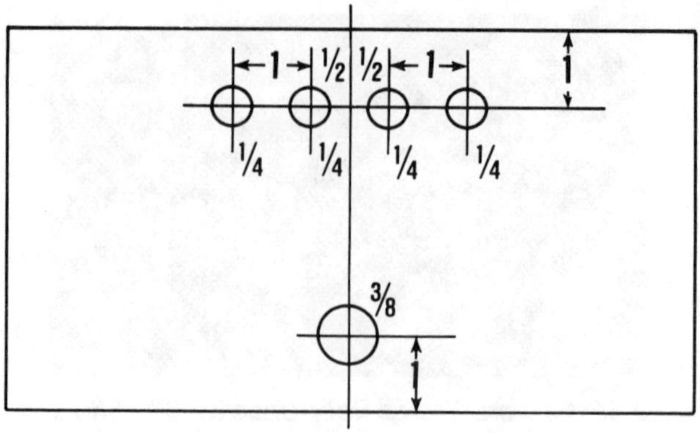

Figure 2-19: Back Panel Drilling

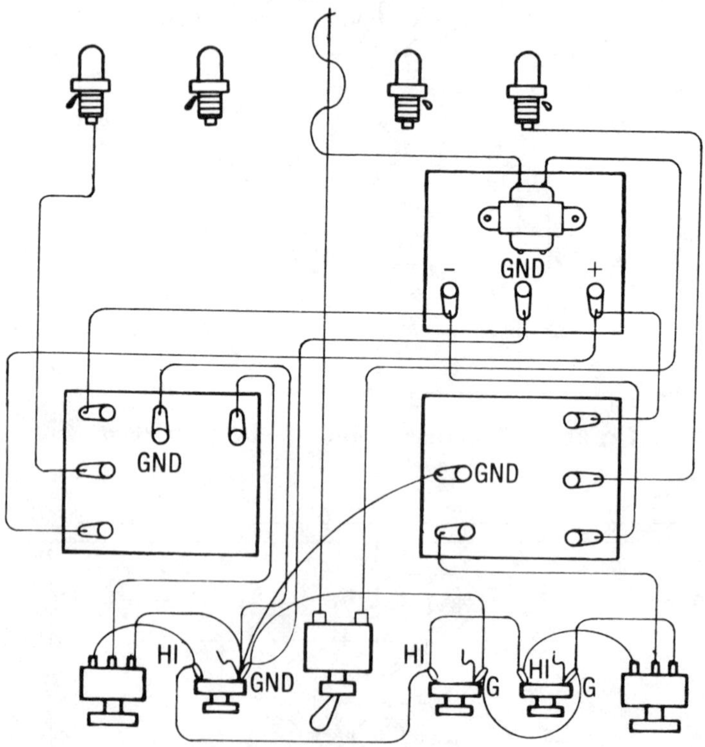

Figure 2-20: Wiring of Components

Figure 2-21: 3 Watt Amplifier, Interior

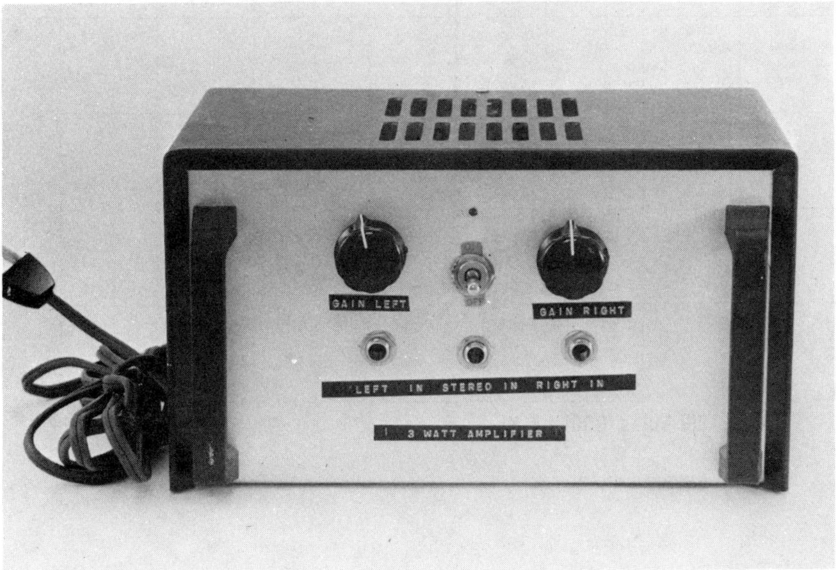

Figure 2-22: Completed 3 Watt Amplifier

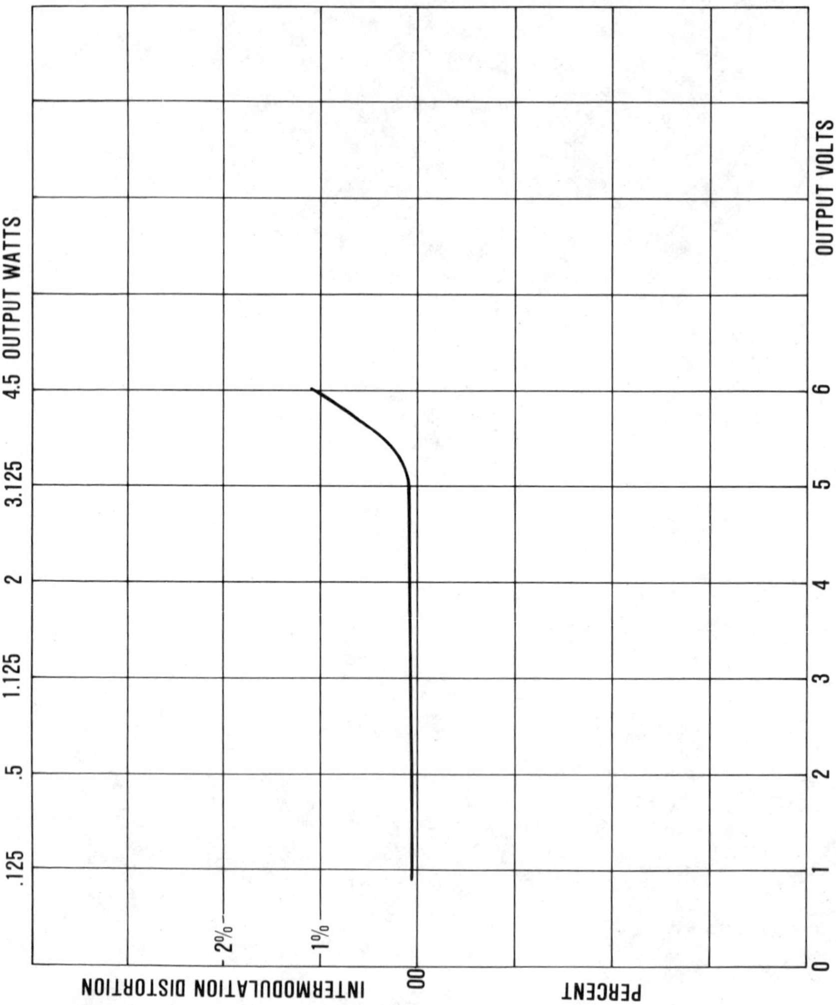

Figure 2-23: Intermodulation Distortion of 3 Watt Amplifier

Adding a Stereo Low Power Amplifier

A second amplifier was built and tested as set forth above. When mounted in the same cabinet also as shown, it provided the stereo amplifier. With the separate gain controls, either amplifier could be operated independently. When used as a stereo system, the separate gain controls provide means for achieving stereo balance.

3

How to Build a 30 Watt Hi Fi Power Amplifier

How the Power Supply Determines Maximum Power

High fidelity transistorized amplifiers seldom use output transformers for a number of reasons. Among these are better fidelity and reduced cost and weight. Without an output transformer, the maximum signal voltage swing across the load is limited by the power supply voltage. Maximum, of course, means the peak swing. The rms voltage is then 0.7 times the peak. Since the standard of comparison is generally the 8 ohm loudspeaker, assume this to be the load impedance.

Table I shows approximately what powers can be obtained from various AC voltage sources in a transformerless power amplifier output to an 8 ohm speaker load.

TABLE I

Power Supply AC Input	No Load DC Volts	Peak AC Volts	rms AC Volts	Power to 8 ohms	Maximum Undistorted Power Output	dB
6.3	8.9	8.9	6.3	4.96	3 watts	0
12.6	17.8	12.6	12.6	19.8	12 watts	6
18	25.4	25.4	18	40.5	25 watts	9
25	35.35	35.35	25	78.1	50 watts	12
35	49.5	49.5	35	153	100 watts	15

How Much Power Is Needed?

The power required in a given system depends on many factors. Let us examine the more important ones and try to arrive at a logical conclusion. We will consider the most general situation, that of the listener in a home environment, either in a house or apartment, and exclude as a special case the "rock" music which is really only suited to wide open spaces.

In talking about high fidelity we can assume that realism is the chief goal. However, that doesn't mean that the realism of first row listening at a symphony concert must be reproduced. The more usual situation is of one sitting half way back in the orchestra. According to the Institute of High Fidelity this would mean a system capable of providing a sound level of 100dB. Also according to IHF, this can be provided in a room of 3000 cu. ft. with a continuous amplifier power of 25 watts per channel in a stereo system. They further state that the figures are for an average room, not too live and not too dead. Now, 3000 cu. ft. describes a room that is 7½ × 20 × 20 feet.

Getting the Most Out of a 25 Watt Amplifier

When discussing amplifier power requirements and capabilities, it is interesting to examine some of the implications. First of all, what is the meaning of power in an amplifier driving a given loudspeaker? Table II shows the relationship between power output to a given loudspeaker and the relative sound levels produced. The figures are again based on what IHF states for a 3000 cu. ft. average room and medium efficiency speakers.

Table II

Amplifier Power output	Sound Pressure Level*
1 watts	85dB
3 watts	90dB
10 watts	95dB
25 watts	100dB
50 watts	102dB
100 watts	105dB
200 watts	108dB
1000 watts	115dB

The numbers in the sound pressure column have been rounded off for easier visualization. For example, 100 watts is actually 6dB greater than 25 watts. The interesting thing to consider is that 3dB is generally considered a just perceptible increase in sound level so that for all practical purposes 100 watts is only two small steps greater than 25 watts in listening level.

Again, the IHF figures 25 watts driving stereo medium efficiency speakers in an average 3000 cu. ft. room is equivalent to listening to a symphony orchestra in a mid-orchestra seat. This should be generally satisfying. However, if one really desires or requires a higher sound level, one way to achieve it without increasing the amplifier power is to use more efficient speakers. Of course such speakers may be more expensive and the cost of more power should be weighed against additional cost of the speakers.

The sound level is as much a function of speaker efficiency as it is of amplifier power. Thus, while 10 times the

*Sound pressure level is the sound pressure expressed in dB with respect to a 0dB reference level. This is the accepted threshold of sound for the average person to midrange frequencies. Generally stated as 1 dyne per square centimeter. A dyne is the force that produces an acceleration of one centimeter per second per second on a mass of one gram. Usual listening levels range from a soft 80dB to a very loud 110dB. Higher levels are dangerous to the ears.

power provides a 10dB increase in sound level, a 10 times increase in speaker efficiency provides the same increase. The graph plotted in Figure 3-11 shows how the sound level in dB increases with speaker efficiency, assuming that with a speaker of 0.1 percent efficiency driven by a 1 watt amplifier produces a sound level of 80dB at 3 feet. There is more about speakers in Chapter 4.

There are several practical factors in favor of the 25 watt amplifier as well as some against much higher power amplifiers. One real factor is cost. In Table III the increase in cost per dB increase in power output is shown for a number of high fidelity stereo amplifiers. The average cost per 3dB increase in output figures is $135. Making a similar comparison for loudspeakers is not as easy since efficiency figures are not generally included in their specifications. In addition to the great increase in cost of high power amplifiers, there is another unfavorable factor and that is safety. If one chooses a moderately priced loudspeaker assuming it will be adequate at 25 watts but drives it with a 100 or even 200 watt amplifier he is definitely endangering the speakers. Most speakers are rated at safe power operating levels, but with a high power amplifier an inadvertent pop from such things as a static burst, or a scratch in a record, may send full power through the speaker and zap it.

Table III

Amplifier Maker	Power Output	Price	dB above 20 watts
Kenwood KA3700	20 w/ch	155	0
Kenwood KA5700	40 w/ch	200	3
Kenwood KA8100	75 w/ch	425	6
Nikko NA550	45 w/ch	230	3
Nikko Alpha II	110 w/ch	460	7.5
Nikko Alpha I	220 w/ch	700	10.5
Phase Linear Dual 500	505 w/ch	1350	14
Harman Kardon hk503	40 w/ch	279	3
Harman Kardon hk505	60 w/ch	399	5

Average cost per 3 dB increase in output = $135

The positive side of using a high power amplifier is that it has "head room." That is, it has reserve power to handle sudden crescendo passages in a lifelike manner. Again, however, this may be at the peril of the speakers.

For the experimenter the practical problems as well as the cost escalate above the 25 watt amplifier level. Components become increasingly difficult to obtain. Construction becomes much more difficult. For all of the above reasons, a 25 watt amplifier represents a very practical compromise.

Your hi fi may actually be superior to sitting in the orchestra, since the microphones used in live programs and recorded programs are much closer to the performers than you can ever be and the result can be much better articulation and clear musical tones. Don't louse up your system by trying to imitate an imperfect concert hall.

Design a 30 Watt Hi Fi Amplifier

In order to increase the power output substantially beyond the 3 watt level, it is obvious that the power supply voltage must be raised. The high performance op amp which proved to be important in a high fidelity amplifier, is rated at ±18 volts maximum supply voltage. With an 18 volt 4 ampere rated power transformer, the no load power supply voltage was 25 volts. The voltage to the op amp was regulated to ±15 volts with two 15 volt Zener diodes.

However, when the simple complementary circuit of the 3 watt amplifier was operated with the 15 volt supply instead of 9 volts, the op amp would drive the output to only about 6 watts. When the output transistors were inverted and a complementary driver added, a full 30 watts was obtained. Figure 3-1 shows the final circuit of the 30 watt amplifier.

The feedback resistor chosen provides full output with an input signal of approximately 0.5 volt rms. In the circuit of the 3 watt amplifier, the output transistors were operated as emitter followers. Thus, the phase of the load voltage is the same as the phase of the op amp output.This means that feedback taken across the load is returned to the inverting input to the amplifier as it would be if taken from the amplifier output terminal.

The output circuit of the 30 watt amplifier includes a complementary driver stage and a complementary power

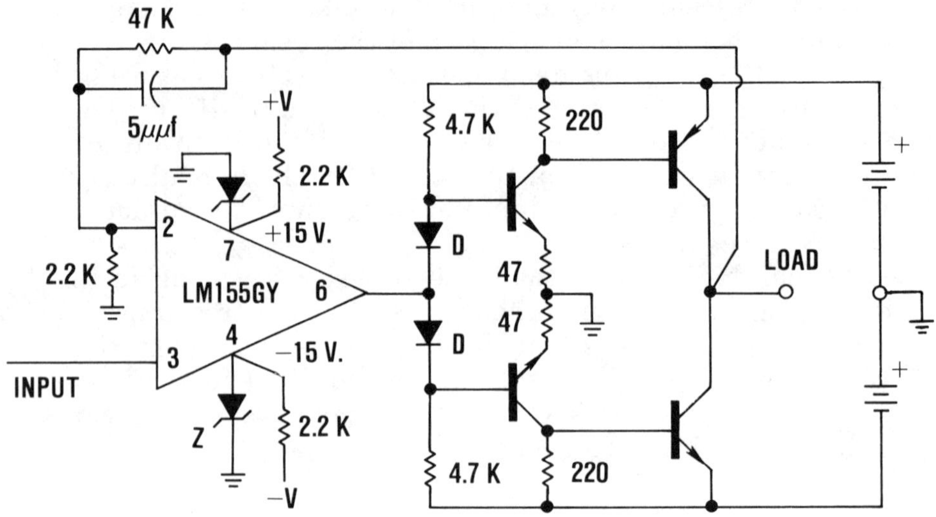

Figure 3-1: 30 Watt Amplifier Circuit

output stage. The outputs of the driver stage are taken from the driver collectors and hence are inverted with respect to the op amp output. The load is also in the collector circuits of the power output transistors and, so too, is inverted. This double inversion puts the phase of the signals across the load back in phase with the op amp output. So again, the feedback taken across the load is returned to the inverting input of the op amp.

Experimentally, the driver load and emitter resistors were chosen to drive the power transistors to somewhat over 30 watts. The driver initial bias is provided by means of diodes supplied with current through 4.7K resistors. In this circuit the diode bias was found feasible, since the initial bias includes the driver and its emitter resistor and drive through the 4.7K resistors is adequate since the drivers require less drive current than did the case of direct power transistor drive in the 3 watt amplifier.

The power supply provided ±25 volts which was too much for the LM1556 op amp and hence it had to be limited as shown with 15 volt Zener diodes.

Experiments were conducted to determine circuit constants in a manner similar to those described in Chapter 2.

Two alternate circuits were evolved, namely, the ones shown in Figures 3-1 and 3-2. The main difference is in the use of diodes as offset controls in Figure 3-1. While the offset with silicon diodes was too high for the circuit of the 3 watt amplifier, they proved satisfactory for the 30 watt amplifier. This was due to a slight difference in the circuits, where in the 30 watt amplifier emitter resistors in the driver stage provided an additional voltage drop. While it was not carried out in the model, some thermal protection can be gained by thermally coupling these offset diodes to their respective power output transistors. Elevated temperatures will tend to reduce the quiescent current in these transistors.

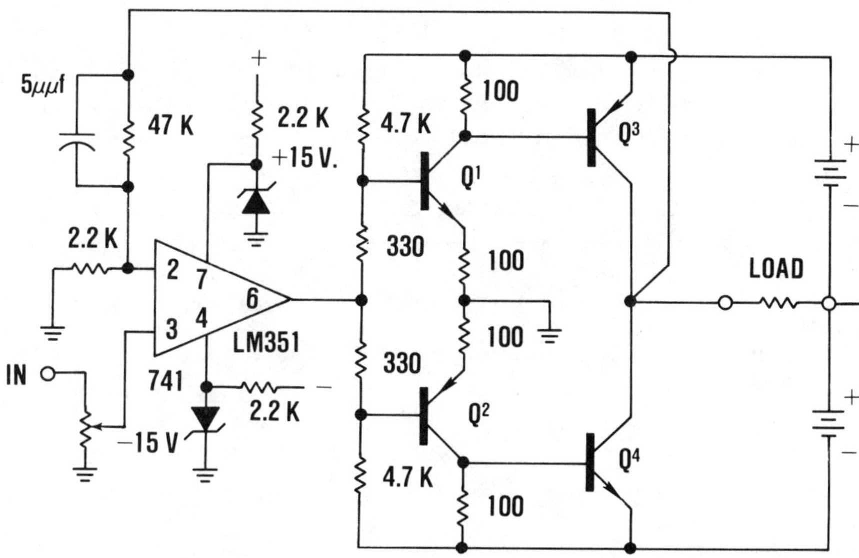

Figure 3-2: Alternate 30 Watt Amplifier

Build the Power Supply for the 30 Watt Amplifier

The power supply for the 30 watt amplifier is a simple device employing an 18 volt 4 ampere line transformer, two diode rectifiers and two 2000 microfarad 35 volt filter capacitors. The circuit is shown in Figure 3-3.

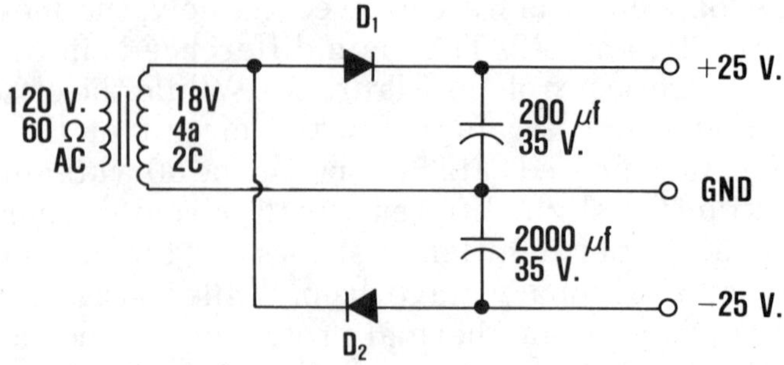

Figure 3-3

Assembly of the power supply was carried out on a piece of perf board 5⅝ by 4¼ inches drilled as shown in Figure 3-4 and assembled as shown in Figure 3-5. Mounting in the cabinet was by using long screws passing through the transformer lugs and the bottom of the cabinet. The capacitors were mounted between two three-point terminal strips.

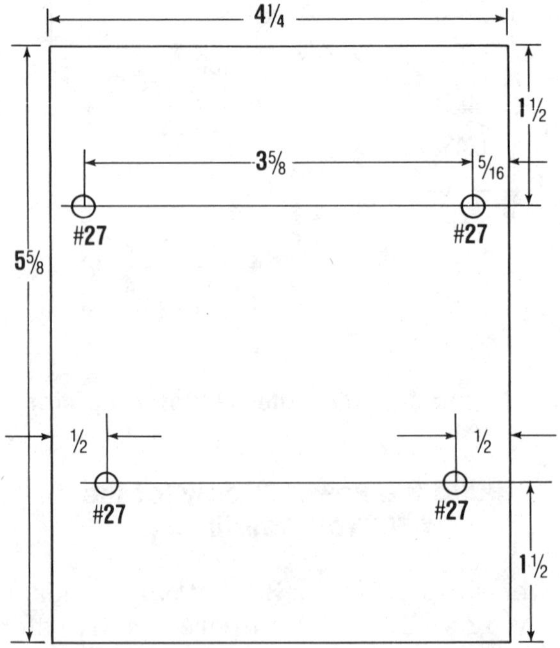

Figure 3-4: Power Supply Board

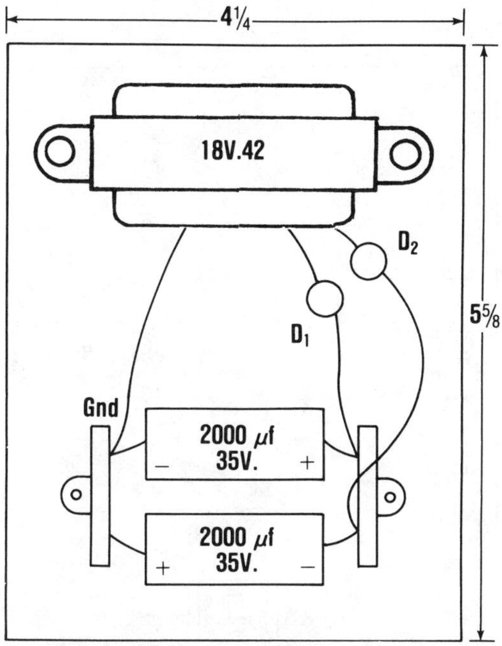

Figure 3-5: Power Supply Assembly

Build the 30 Watt Amplifier

The op amp and driver transistors were mounted on a 3 by 2⅜ inch piece of perf board and wired as shown in Figure 3-6. The dotted leads are on the under side of the board while the components and solid leads are on the upper side. The brackets holding the heat sinks and the driver transistors are also shown. Color coded leads to other components are shown and labeled. The PNP and NPN designations are to the respective terminals of the output power transistors. This circuit board was mounted by means of an angle bracket to the heat sink board within the amplifier cabinet to prevent the underboard wires from shorting on the bottom of the cabinet.

The power output transistors were TO-3 can type and were mounted on flanged heat sinks. These heat sinks were mounted on a piece of phenolic board drilled as shown in Figure 3-7 which, in turn, was mounted vertically in the amplifier cabinet.

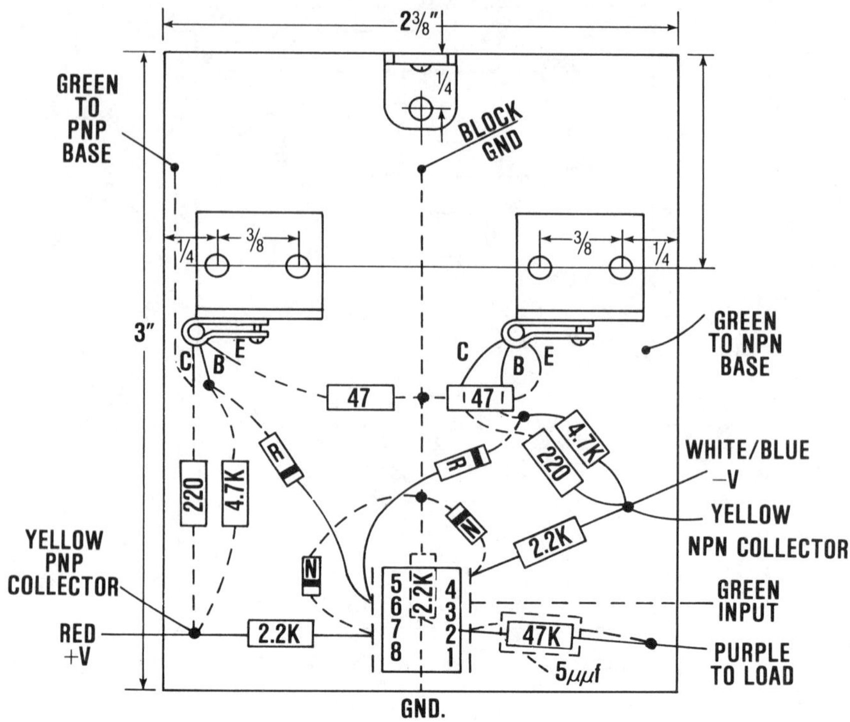

Figure 3-6: 30 Watt Amplifier Circuit Board

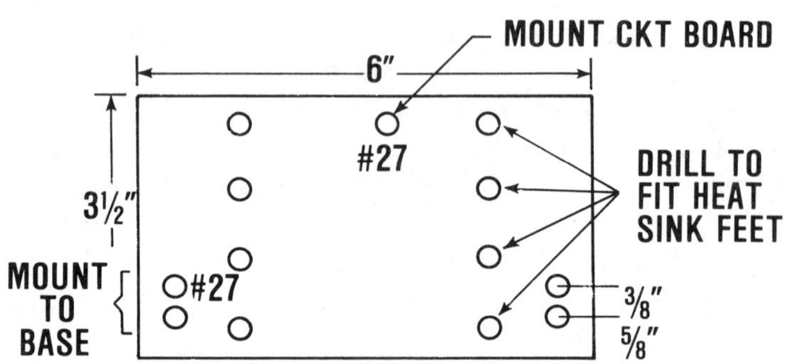

Figure 3-7: Heat Sink Mounting

Suggested Parts for the 30 Watt Amplifier

Resistors—¼ watt 5 or 10% carbon
Zener diodes—12 or 15 volt
Op amp—LM1556V or LM351
Drivers—Sylvania ECG192 and ECG193
Power transistors—Sylvania ECG280 and ECG281
Rectifiers (D_1 & D_2 Fig. 3-1)—2 amp 100 v. silicon

Enclosing the 30 Watt Amplifier

The 30 watt amplifier and its power supply were mounted in a Radio Shack Cat. No. 270-0270 cabinet. The front panel was drilled as shown in Figure 3-8. The finished amplifier is shown in the photographs Figures 3-9 (outside), 3-10A, 3-10B and 3-10C (inside).

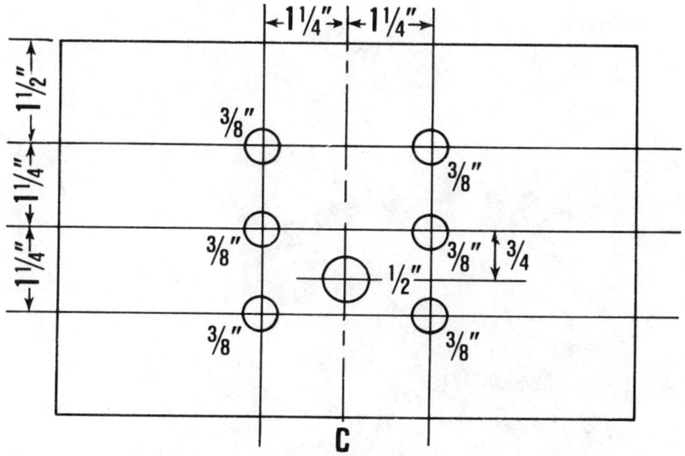

Figure 3-8: Front Panel Drilling

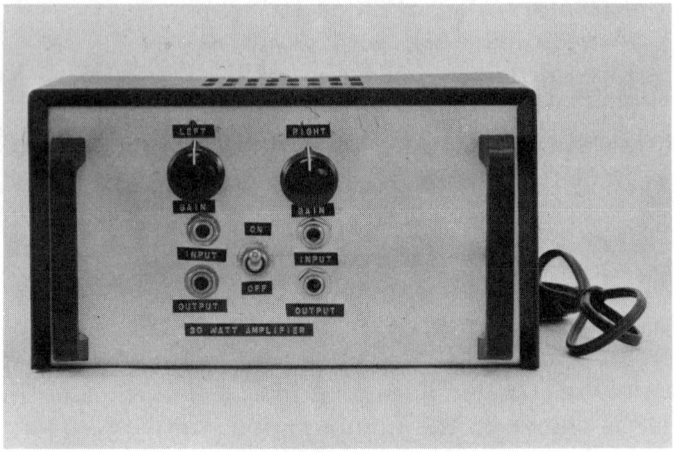

Figure 3-9: 30 Watt Amplifier

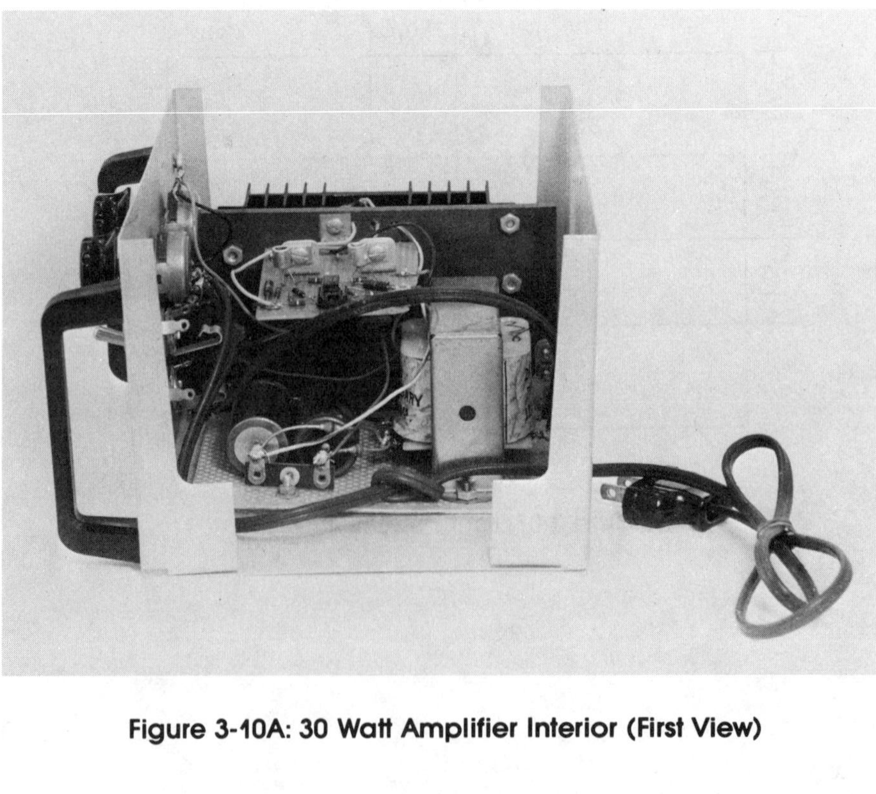

Figure 3-10A: 30 Watt Amplifier Interior (First View)

Figure 3-10B: 30 Watt Amplifier Interior (Second View)

Figure 3-10C: 30 Watt Amplifier Interior (Third View)

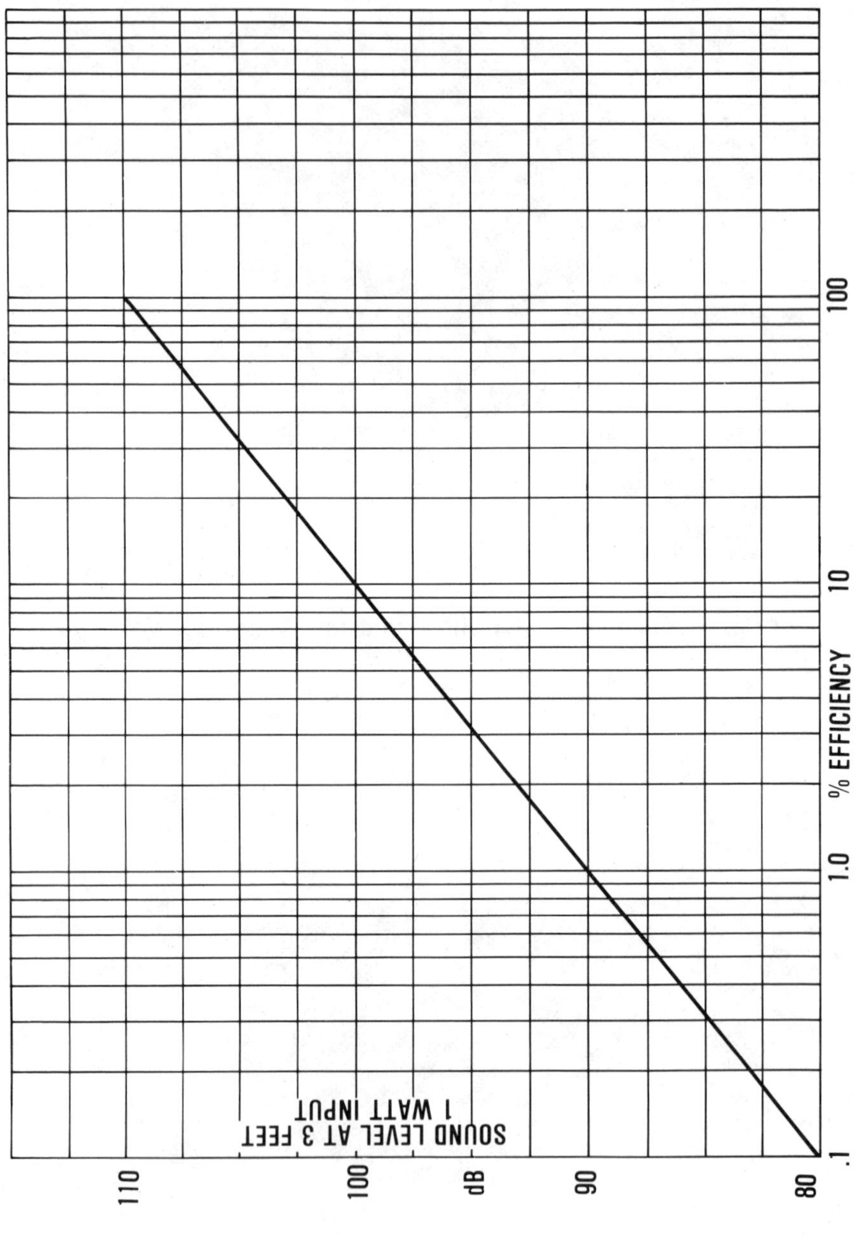
SOUND LEVEL AT 3 FEET
1 WATT INPUT
dB
110
100
90
80
% EFFICIENCY
.1
1.0
10
100

Figure 3-11

4

Building a Full Size Hi Fi Loudspeaker

Examining Loudspeaker Problems

Loudspeakers are the weak link in the high fidelity system. Amplifiers are the strongest link since they can be made to have undetectably low orders of distortion. While other components such as microphones, recorders and reproducers may have measurable imperfections, they are still ahead of the final link, the loudspeaker.

What Is the Ideal Loudspeaker?

The ideal loudspeaker converts the audio input voltage applied to it into precisely corresponding variations in sound pressure throughout the listening area. It has a perfectly flat frequency response over the entire audible frequency range. It produces only undetectable distortion.

There are many speakers on the market. Over 60 reputable manufacturers offer highly-engineered products ranging in price from less than $50 to over $1500 in wide-ranging attempts to produce the perfect speaker. While published specifications give some clues as to the differences between them, it is impossible to describe in words how close to reality the output of these speakers actually comes. Thus, it is advisable to make a listening test before making a final decision. However, here again there are problems. One is: how expert is one in judging quality? Another is: a listening test made in some manufacturer's show room does not reproduce one's own listening room environment.

The Listening Room is Important

The listening room (the room where you are going to enjoy your high fidelity system) is a major factor in high fidelity reproduction as well as the placement of speakers in that room. In any enclosed area, other than an anechoic chamber, reverberation has an important influence on loudspeaker performance. A live room, one with bare walls, reenforces sounds and often emphasizes certain bands of frequencies. This tendency can be at least partially countered by equalizers. A dead room on the other hand requires more power to fill adequately. Again, speaker placement may have a dramatic effect on realism. All these matters will be discussed in detail below.

The Significance of Speaker Enclosures

A bare loudspeaker exhibits a very poor frequency response especially at the low frequency end. This is because the sound radiation from the rear of the speaker is not blocked and as a result cancels much of the front side radiation. When a speaker is mounted on a baffle, the rear-side radiation is blocked more or less depending on the size of the baffle. An infinite baffle blocks it all, but infinite baffles are generally impractical. Mounting a speaker in a door or wall is an example of an essentially infinite baffle but is not particularly convenient and is limiting in other ways. The practical answer is an enclosure, of which there are many types.

Enclosures have one outstanding advantage—they can be moved around.

How to Provide Wide Frequency Coverage

No one dynamic speaker is capable of covering the entire audible frequency spectrum with a flat output. In order to go down to the lowest frequencies, say to around 20Hz, a speaker of 12 inches in diameter is required. However, such a speaker in no way can reproduce the high end, say to 20kHz. Most speaker systems covering the full range employ a low frequency speaker (woofer) a midrange speaker, and a high frequency speaker (tweeter). Fortunately the three can be housed in the same enclosure with satisfactory results.

With three speakers the frequency spectrum from 20Hz to 20kHz is divided into three bands. Each speaker is confined to its optimum band by means of cross-over networks (see p. 83). This technique delivers full power to each speaker for its own frequency band. This conserves power, protects the speakers from being overloaded with signals other than its own designated band and reduces distortion. For example, using cross-over points of say 700Hz and 7kHz causes the woofer to operate essentially from 20Hz to 700Hz, the midrange speaker from 700Hz to 7kHz and the tweeter from 7kHz to its upper limit of 20kHz or so.

Harmonic Distortion in Loudspeakers

Distortion in loudspeakers is not well documented. There are probably two important reasons for this. One is that distortion is a complex quantity varying with applied frequency and amplitude of the drive. It also is a function of the internal impedance of the driving amplifier, a factor over which the speaker manufacturer has no control. Perhaps the most important fact is that distortion figures for loudspeakers do not compare very favorably with the distortion found in good amplifiers.

All of the above leads to the conclusion that in order to get the most out of a high fidelity system great care should be exercised in selecting a loudspeaker.

Selecting Loudspeakers for Experimenting

Assume the object of this chapter is to build a good high fidelity speaker including a woofer, midrange and tweeter. The following three speakers are a good starting point:

1. Radio Shack No. 40-1007 12″ woofer rated at 22Hz to 550 Hz with 50 watt rms power handling capability.
2. Radio Shack No. 40-1005 midrange rated at 800Hz to 15,000Hz with 60 watt rms power handling capability.
3. Radio Shack No. 40-1276 high frequency dome tweeter rated at 6000Hz to 18,000Hz with 40 watt rms power handling capacity or,
   (In place of 3 above) Radio Shack No. 40-1310 super tweeter rated at 5000Hz to 40,000Hz with 50 watt power handling capacity.
4. Radio Shack No. 40-1294 3-way crossover which crosses over at 1000Hz and 6000Hz.

Equivalent speakers are available from other sources. As stated above, the speaker system is the one limiting factor in any high fidelity system worthy of the name. For this reason speakers should be selected with great care. Radio Shack speakers are recommended on the basis of acceptable performance and generally wide availability.

Choosing the Bass-Reflex Enclosure

There are many types of speaker enclosures in use. They range all the way from closed boxes (air suspension) to complicated folded corner horns. While the horn is by far the most efficient in converting electrical currents to sound pressure, they are complicated and difficult to build. The simplest and least efficient is the air suspension. The closed box, if vented properly, becomes the second most popular type, namely, the bass-reflex. The bass-reflex speaker smooths the low frequency response and increases efficiency. Taking into consideration efficiency, frequency coverage and the complexity of construction, the bass-reflex has been chosen for experiments and the loudspeaker Project.

Measure Free Air Characteristics

In order to fully appreciate the importance of baffles and enclosures, it is interesting to measure the woofer characteristics in free air. Start by hanging the speaker in a space away from any walls or obstructions. Connect an audio generator to the speaker through a 100 ohm resistor and measure the voltage across the speaker as the frequency is changed as illustrated in Figure 4-1. The speaker impedance is closely equal to $100e_2/(e_1\text{-}e_2)$. The woofer used in this project was so measured and the results are plotted in Figure 4-2.

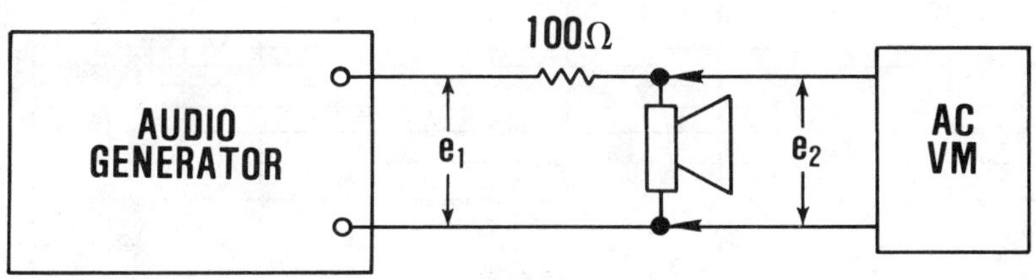

Figure 4-1: Audio Generator

If this speaker were to be driven by a high output impedance amplifier, the sharp peak at 35Hz would be very pronounced.

However, another run was made driving the speaker by means of the 3 watt amplifier described in Chapter 2. The voltage across the speaker and the sound level were measured at close range (18 inches to avoid room reflections). The results are also plotted in Figure 4-3. The voltage over most of the range was within plus or minus 1 dB. The sound level, of course, fell off rapidly below 500Hz due to lack of a baffle. The latter curve also shows why the woofer must be supplemented with one or more high frequency speakers as its responses fall rapidly above 2000 cycles.

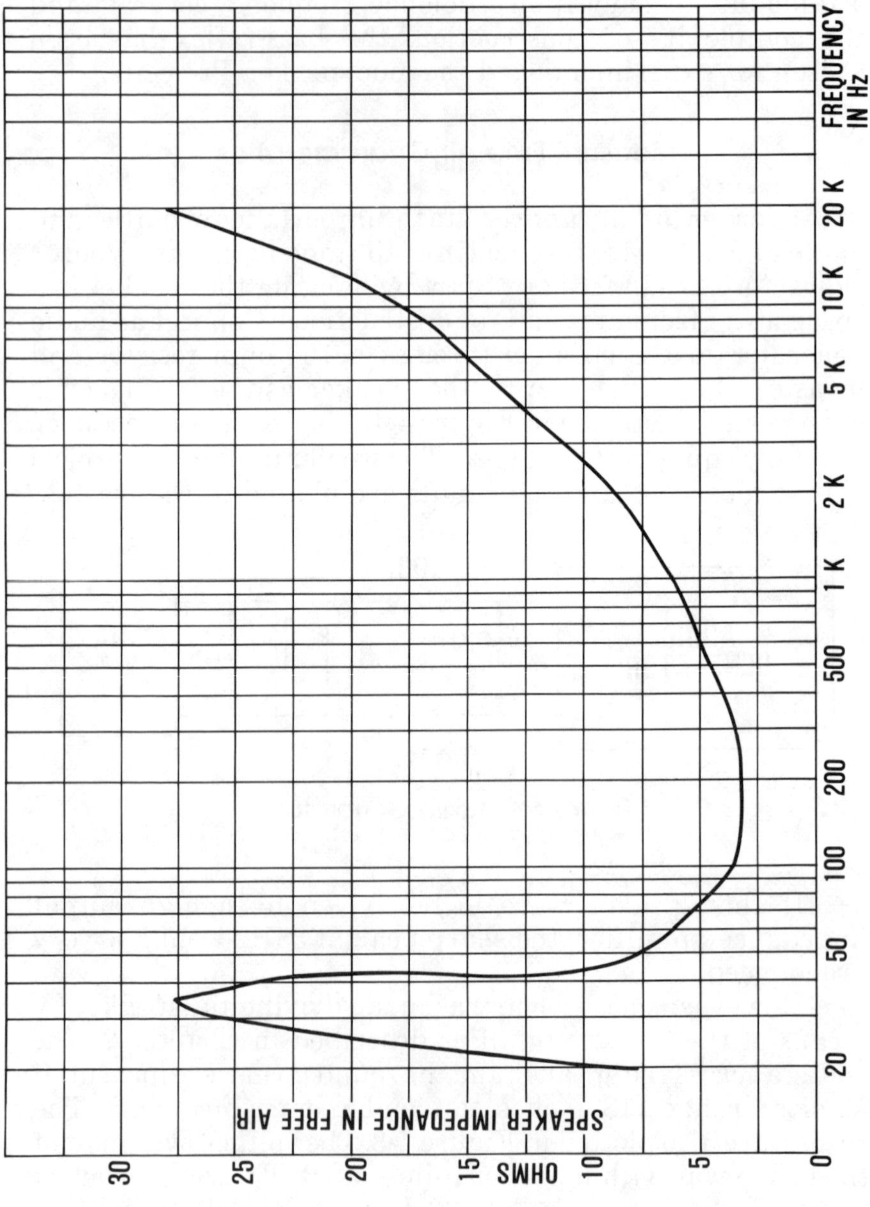
SPEAKER IMPEDANCE IN FREE AIR
OHMS
0
5
10
15
20
25
30
FREQUENCY IN Hz
20
50
100
200
500
1 K
2 K
5 K
10 K
20 K

Figure 4-2

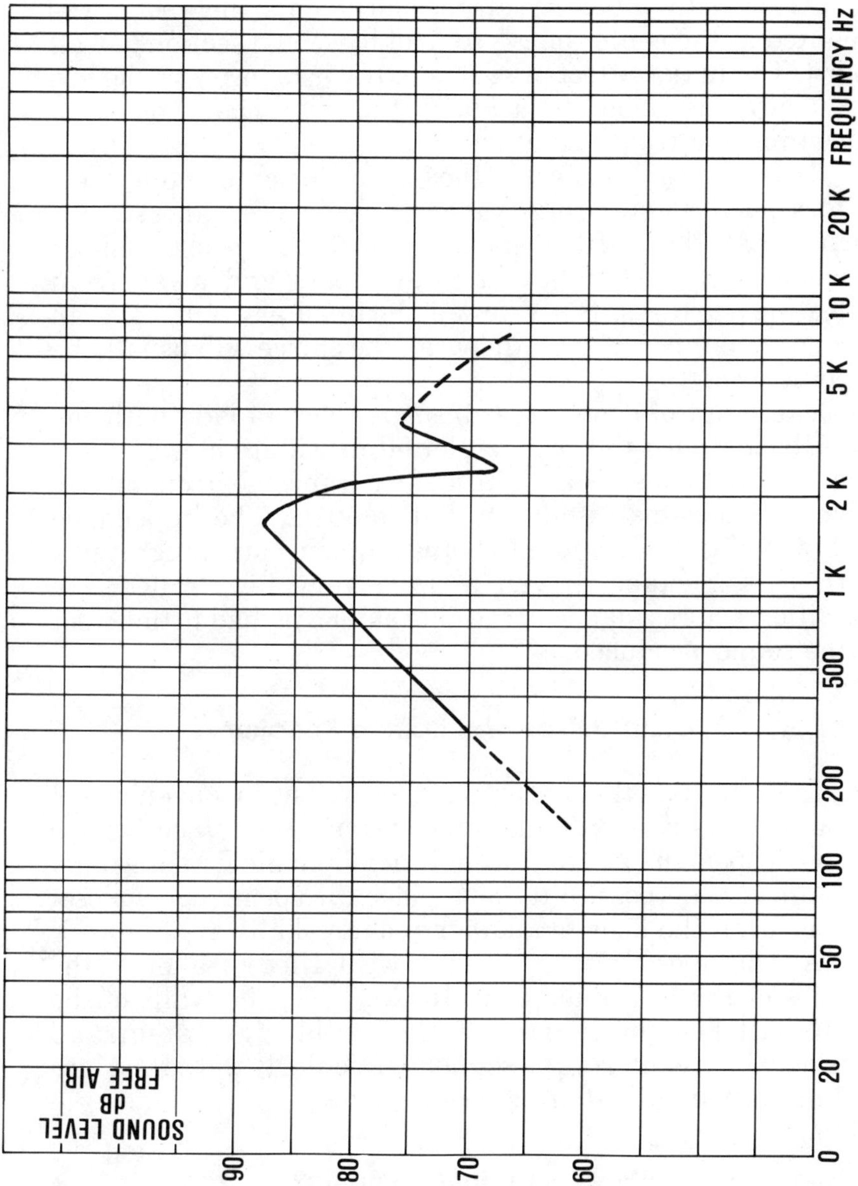
FREQUENCY Hz
20 K
10 K
5 K
2 K
1 K
500
200
100
50
20
0
90
80
70
60
SOUND LEVEL
dB
FREE AIR

Figure 4-3

Mount Woofer on a Flat Baffle

After making the free air measurements, the woofer was mounted at the center of a 4 × 6 foot by ½ inch section of wall board. The major effect of such a baffle is to improve the low frequency radiation by keeping the rear radiation from cancelling the front radiation.

To evaluate the effect of the baffle, do not introduce new factors such as standing waves or room resonances. This means that the baffled speaker must be checked in an anechoic (dead) chamber or out-of-doors and away from reflecting surfaces. Since few experimenters will have an anechoic chamber at their disposal, the choice will usually be out-of-doors.

These out-of-doors measurements were made with the 4 × 6 baffle with the results plotted in Figure 4-4. The low frequency response was significantly improved from 500Hz down and became essentially flat to 50Hz. The impedance peak was slightly lowered in both frequency and magnitude. With this speaker driven from a very low impedance amplifier, the resonance in the characteristic had little effect on the sound produced.

Possible Uses of a Baffled Speaker

In most cases the baffled speaker will be inconvenient on account of its size, particularly when stereo is an objective and two such speakers are required. If monophonic listening only is required, one possibility is to place the speaker in a closet door or even the door to another room or hallway.

Another possibility is to place two baffled speakers in the corners of the listening room. In this case, the walls of the room will become a part of the radiating system and considerably reenforce the sound, particularly the bass. More will be said about listening rooms below.

What Is an Infinite Baffle?

A true infinite baffle is just what the name implies, a baffle of infinitely large dimensions. Naturally, this is

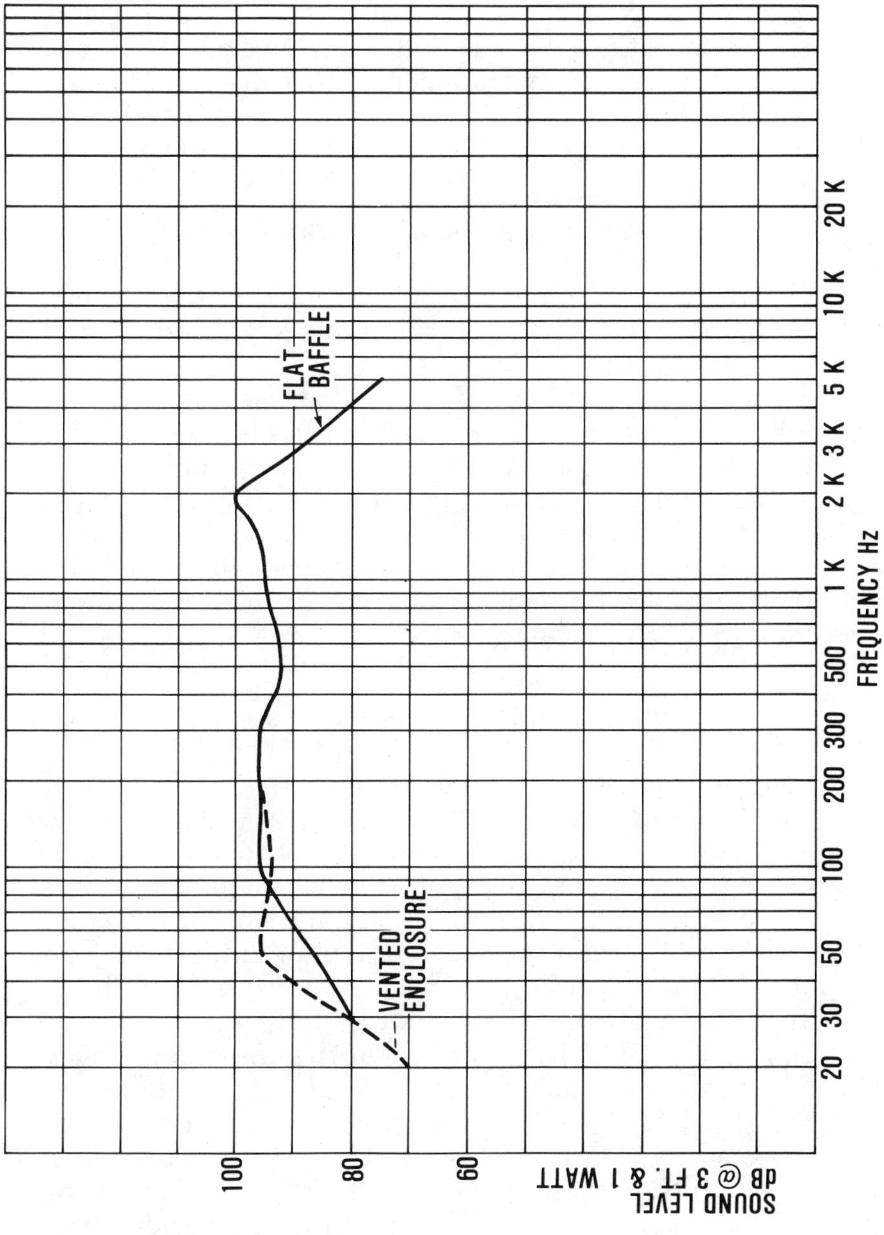

Figure 4-4: Woofer in 4′ × 6′ Baffle Out-of-Doors @ 1 Watt

impractical but infinite baffles can be used effectively in a number of ways. A speaker in the wall or door of a room is one example. A speaker in the door of a good sized closet is another. The most usual is a closed box. However, the closed box is not a true infinite baffle since it introduces new factors into the picture.

How Does a Closed Box Affect Speaker Performance?

Two things are assumed at the outset; one is that the box is well damped with sound absorbing material inside it; and the other is that it is of adequate size.

First, the damped box will load the speaker and provide a resilient air mass behind the speaker. Second, the box will affect the natural resonant frequency of the speaker. It will also substantially decrease the acoustic efficiency of the speaker. On the positive side, it will make satisfactory reproduction possible in a small container. So-called bookshelf speakers are typical examples of often surprisingly good reproduction in a small space.

Build a Full-Size Hi Fi Speaker

There is almost no limit on the size and complexity of high fidelity speakers. Likewise store prices range from around $100 to many hundreds of dollars. The price for size, complexity and cost goes up rapidly as one reaches for the last ounce of perfection. On the other hand, none of these factors guarantee more than what a given manufacturer can produce. While scientific measurements are not infallible, even they will show that there is still a long way to go.

Considering this, in view of all of the effort and money which has gone into loudspeaker development over the years, it seems rather futile for the average experimenter to hope to produce a nearly perfect speaker. But he or she can produce a very good and perhaps outstanding speaker. One to be especially proud of for having made it himself or herself.

As in all experiments and projects, we will try to combine the practical with the esoteric. We will use proven principles, but practical ways of carrying them out. First, we will make a closed box speaker of generous dimensions with a 12 inch woofer, a 5 inch midrange and a 3½ inch tweeter.

Constructing the Enclosure

A satisfactory, popular and readily available material for the enclosure is ¾ inch plywood. A standard size for such wood is 4 × 8 feet. A little calculating shows that one can build a 36″ × 24″ × 24½″ speaker from one piece of plywood with only about 2 percent waste. Figure 4-5 is a dimensional drawing of the basic enclosure. Figure 4-6 shows how the plywood is to be cut. Eight cuts does it all.

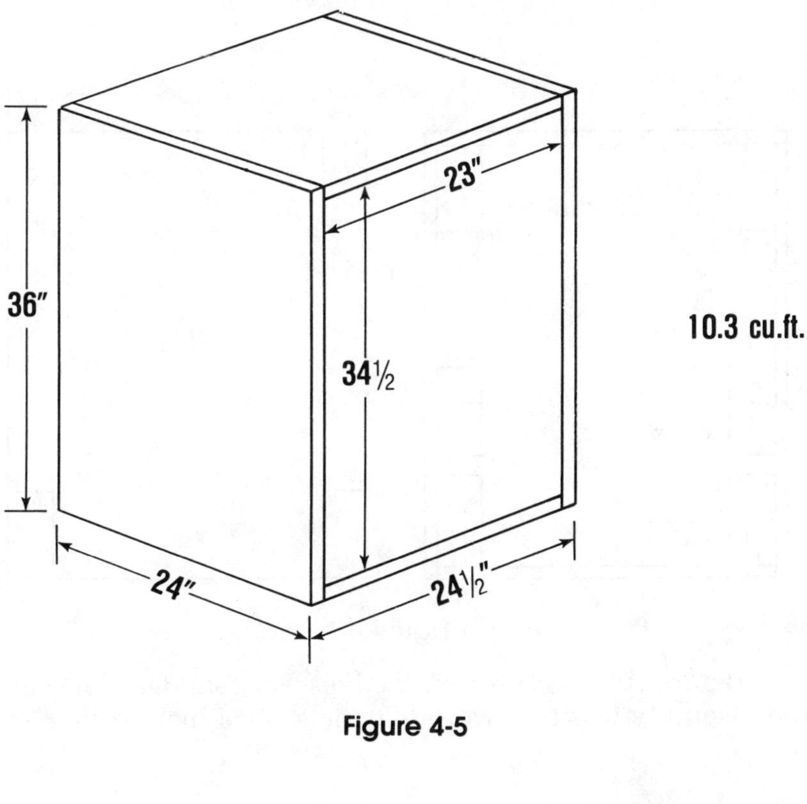

Figure 4-5

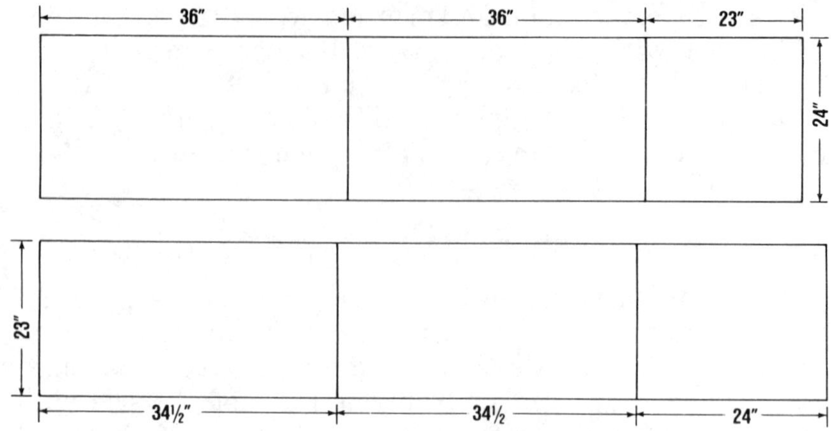

Figure 4-6: 4 × 8 × ¾ Plywood Cut for Full Size Speaker

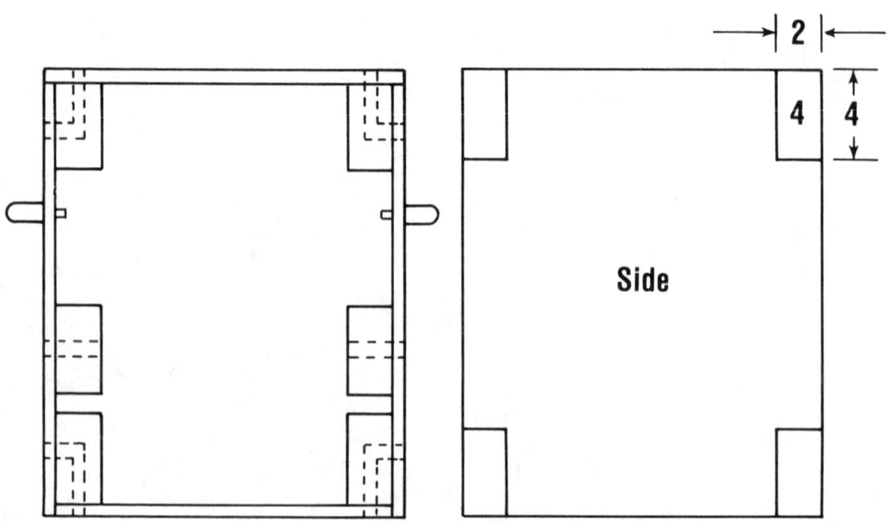

Figure 4-7

Before final assembly, all edges were sanded. The key to the assembly was the two side panels. Two-inch by four-inch blocks were glued in the four corners of these side panels as shown in Figure 4-7. Then the panels were securely nailed to the blocks. Top and bottom panels were mounted using 10×2 flat head wood screws. Clearance holes were drilled in the top and bottom panels and the outside ends were counter sunk. The panels were aligned on the side panels and small pilot holes drilled in the blocks. The top and bottom panels were mounted with the mounting screws securely turned. The back panel was mounted in the same way resulting in a rugged and accurate rectangular enclosure. Two carrying handles were mounted, one on each side panel. All cracks around the joints were filled with plastic wood. The cabinet front view is also shown in Figure 4-7.

Stiffening the Enclosure

The enclosure was stiffened by means of 2 × 2 × 24 inch wood strips glued and screwed on diagonals to the four sides and back of the enclosure. To facilitate moving the speaker from place to place in order to test the sound, heavy castors were mounted on the bottom in addition to the two handles on the sides of the enclosure. These proved to be a great help, especially when taking the speaker out-of-doors.

Adding Sound-Deadening Material

It is highly recommended that sound-absorbing material be added inside the enclosure. How much is not an easily answerable question, but it should be a generous amount. Figerglass is probably the most practical material to use from the standpoints of availability and effectiveness.

In the speaker cabinet described, I mounted built-up bats to 2 inches thickness, 14 × 27 inches on two sides and the back and 4 inches thick by 18 × 18 inches on the top. The bats were secured by means of small nails passed through squares of strong but thin fiberboard. There is no easy way to evaluate this sound absorbing treatment but it looked like a generous application.

Venting the Enclosure

I decided to use a vent having an area of 95 square inches, a size suitable for a 12 inch speaker. In order to do this, it was necessary to mount the front panel of the enclosure up from the bottom panel by 4.2 inches. This was done by adding two additional blocks as shown in Figure 4-7 and Figure 4-16 and screwing the front panel to these and the top two blocks.

Mounting the Woofer

First, an 11 inch diameter hole was cut with a hand operated jigsaw centered 12 inches above the lower edge of the front panel as shown in Figure 4-8. Four mounting holes were spotted through the speaker frame and the woofer was securely mounted with thru-bolts.

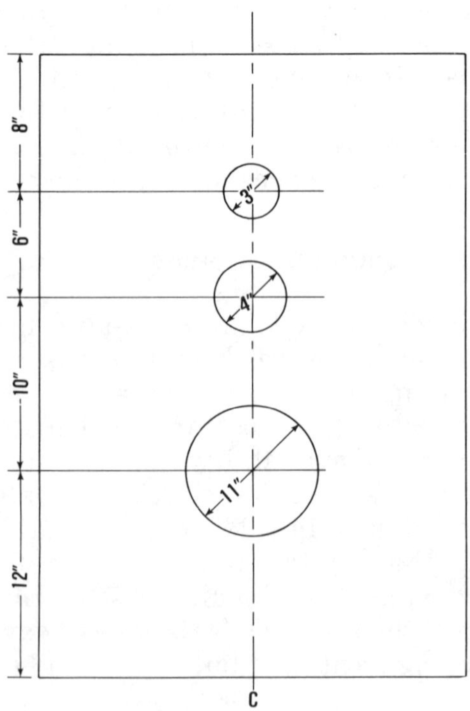

Figure 4-8: Mounting Holes to Match Speaker Holes

Testing the Woofer in the Enclosure

The impedance of the woofer mounted in the vented enclosure was measured as described above. The result showed a slight decrease in the frequency of the main resonant peak and a second peak at a somewhat higher frequency (see Figure 4-9). We will see later that these peaks have little effect on the sound output at low frequencies when the speaker is driven by an amplifier having a high damping factor.

The enclosed woofer was tested outdoors under the same conditions under which the baffled speaker was tested. The results are plotted in Figure 4-4 along with the baffled speaker results. It is interesting to note a substantial boost at 50Hz but a drop below 30Hz. One seldom finds program material with which a difference could be noted between these two conditions. Averaging the sound level between 500Hz and 1kHz indicates a speaker efficiency of 1.75 percent—a rather average figure.

Add Midrange and Tweeter

The next step in the full size speaker construction is to cut the 4 inch and 3 inch holes as shown in Figure 4-8 and to mount the midrange and tweeter speakers. A three speaker system such as this is called a "three way speaker." Special means must be taken to operate a three way speaker efficiently.

How to Use Cross-Over Networks

The typical three way speaker system employs a woofer, a midrange speaker and a tweeter. Each of these speakers is designed to most effectively cover a specific band of frequencies. If the three speakers were simply connected in parallel or in series, several problems would present themselves. In the first place, the resulting impedance would be too low in the parallel case, and too high in the series case for efficient amplifier operation. These connections would also waste power because for any given band of frequencies,

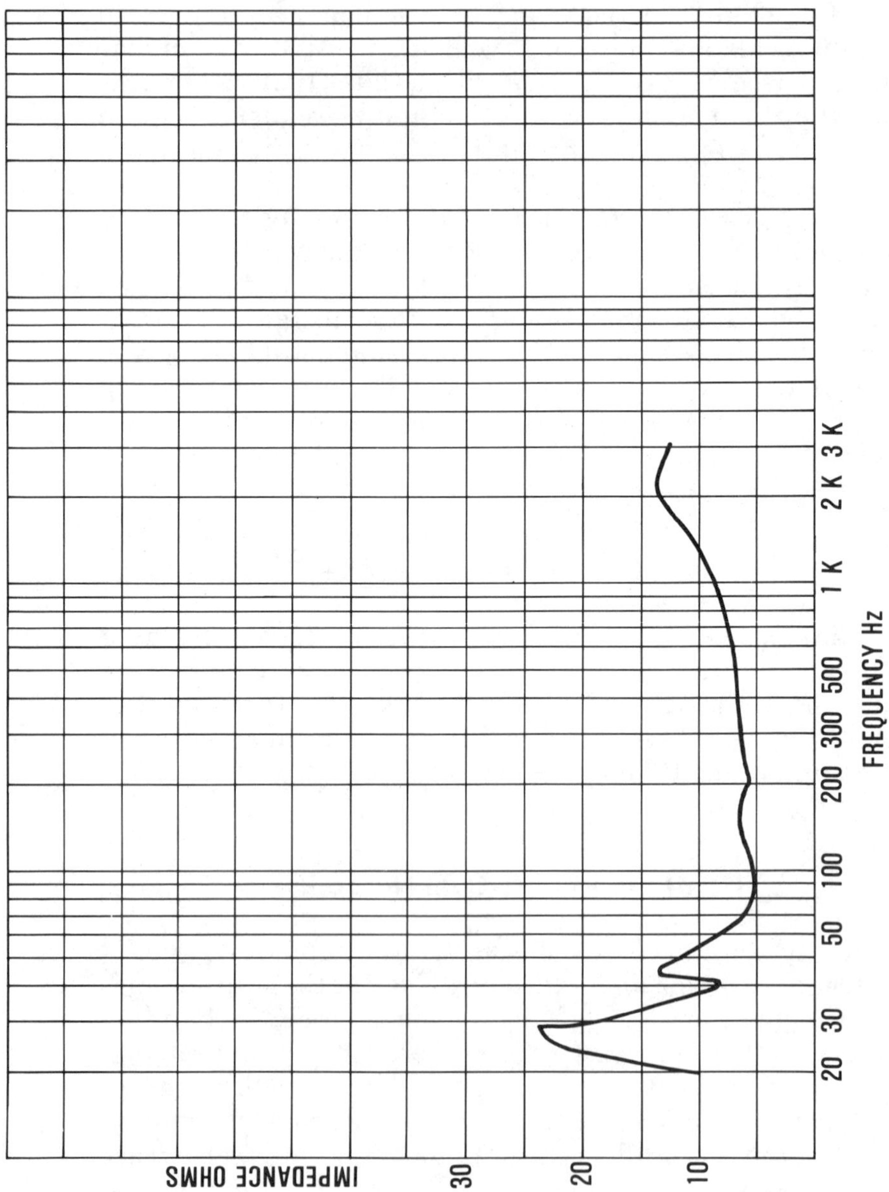

Figure 4-9: Impedance of Woofer in Vented Enclosure

two out of the three speakers would consume power without producing any useful acoustical output. Also a speaker that consumes power but does not provide sound output may be burned out.

The accepted solution to the problem is to use cross-over networks. These networks selectively apply power to a given speaker over a predetermined band of frequencies. These frequencies are chosen to be the optimum response band of each speaker. The networks are selective so that each speaker operates substantially alone except for in-between bands called the cross-over region. In the cross-over region, one speaker is being phased out and another phased in.

A two way speaker system requires only a low-pass filter for the low frequency speaker and a high-pass filter for the high frequency speaker. The two filters are designed to cross-over at a common frequency called the cross-over frequency. That is, the low-pass filter is designed to cut off at, say, 550Hz and the high-pass filter is designed to cut off at the same frequency. Figure 4-10 shows a low frequency speaker and a high frequency speaker coupled to an amplifier through a low-pass filter and a high-pass filter respectively.

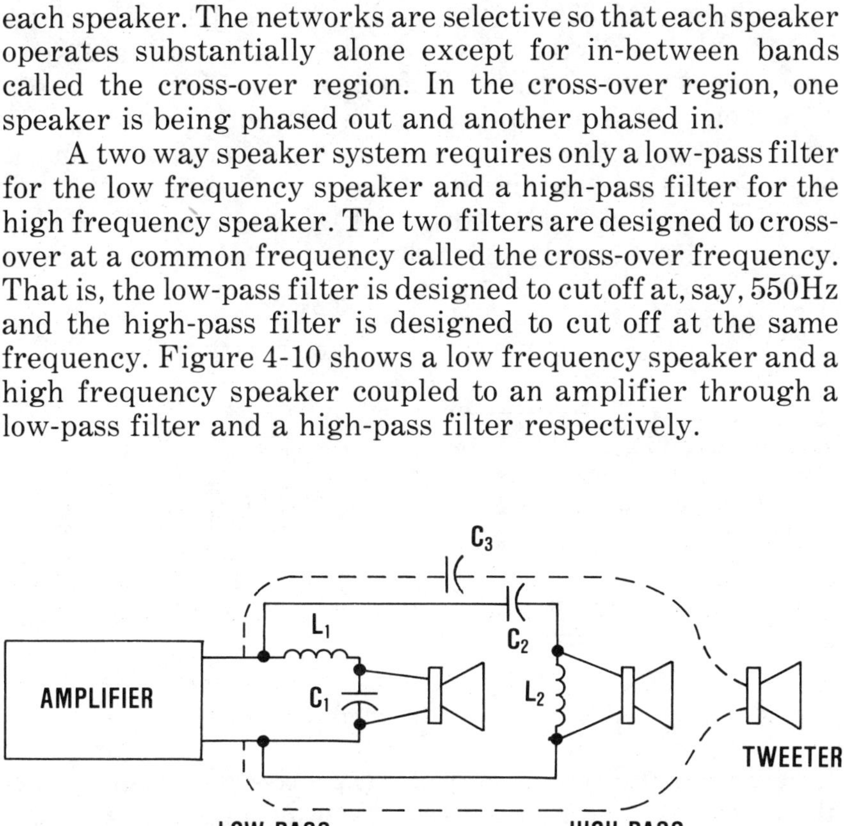

LOW PASS
$C_1 = 36\ \mu f$
$L_1 = 2.3$ mh

HIGH PASS
$L_2 = 1.15$ mbh
$C_2 = 18\ \mu f$

TWEETER $C_3 = 3\ \mu f$

Figure 4-10

Design Low-Pass and High-Pass Filters

While the theory of cross-over networks can lead to some very complex concepts and filters to implement these concepts, we will deal with some basics which provide practical solutions. The most commonly used low-pass filter is the basic half section type as shown in Figure 4-10A. The formulas for L and C are shown in the Figure 4-10B for a cut-off frequency of 550Hz. Similarly, Figure 4-10B shows the compatible half section high-pass filter again with values of L and C worked out for a cut-off frequency of 550Hz. Figure 4-10A shows how these filters are used to couple low and high frequency response speakers to an amplifier. The R_o in the formulas is the rated impedance of the speakers.

Choosing the 550Hz cross-over frequency and the 8 ohm impedance, the L and C for the low-pass filter were calculated

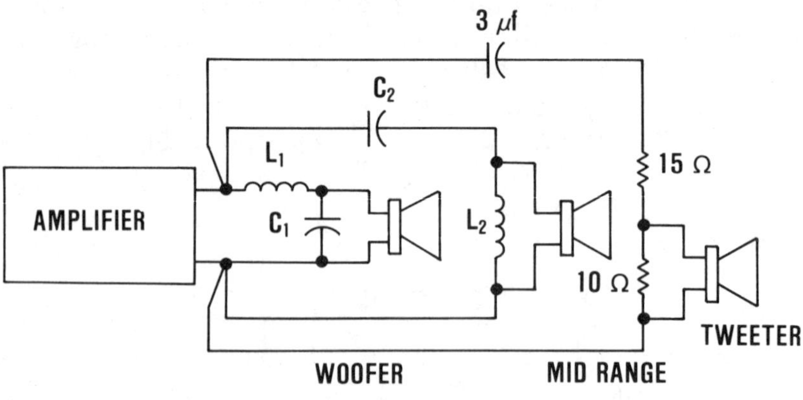

Figure 4-10A

to be 2.3 mh and 36 microfarads respectively. The half values then would be 1.15 mh and 18 microfarads. Two non-polarized capacitors (Radio Shack No. 272-999) in parallel provided 20 microfarads, close enough, for the capacitor. The 1.15 mh inductor, however, is not easy to come by. This inductor must obviously be a very low resistance inductor having low losses at low audio frequencies since it is to be used in series with an 8 ohm speaker. Finding no ready source for the inductor, I decided to make it.

The finished inductor is shown in Figure 4-11. A form was made using 11/32 inch diameter powdered iron core 2¾

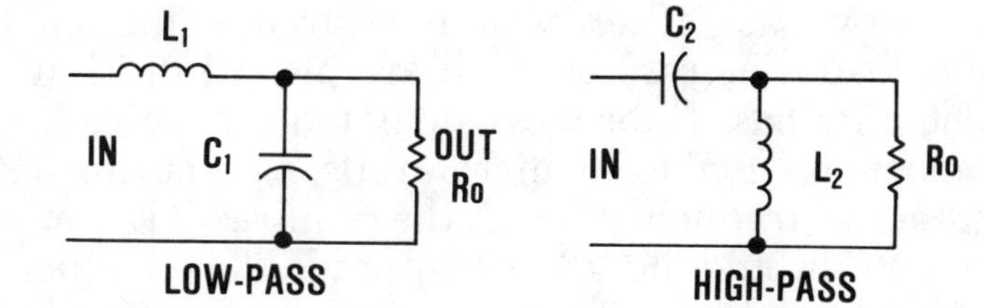

$$L = \frac{Ro}{Wc} = \frac{8}{3454} = 2.3 \text{ mh}$$

fc = 550 Hz Wc = 3454

$$C = \frac{1}{WoRo} = 36\ \mu f$$

$$L = \frac{Ro}{2Wc} = 1.15 \text{ mh}$$

fc = 550 Hz

$$C = \frac{1}{2WcRo} = 18\ \mu f$$

Figure 4-10B

inches long, the core of a portable radio antenna coil. Two end pieces 2 inches square of thin fiber board were cemented to the core. Then 200 turns of No. 18 enameled copper wire were wound on the form. The inductance was checked by finding the resonant frequency with a 0.5 microfarad capacitor. It turned out to be 6.5kHz, indicating an inductance of 1.2 mh. The Q was high, indicating low losses. The DC resistance measured 0.3 ohm, a very acceptable figure.

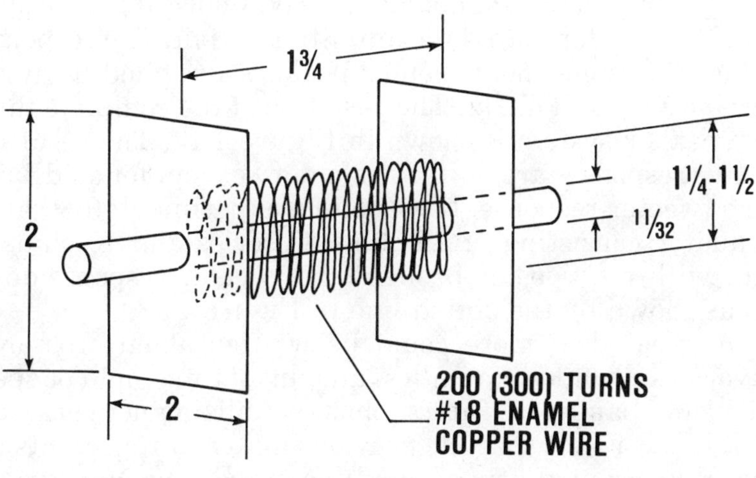

Figure 4-11

The low-pass filter was measured using an 8 ohm resistive load. The results in dB are plotted in Figure 4-12.

The high-pass filter was calculated to require L equal to 1.15 mh and C equal to 18 microfarads, again using 550Hz as the cross-over frequency. Thus, the required 2 L would be 2.3 mh and 2C would be 36 microfarads. The inductor was provided by winding 100 more turns on the low-pass inductor. It measured 2.8 mh, a little high, so by removing a few turns it was trimmed to the required 2.3 mh. The capacitor was formed by using the two 10 mfd nonpolarized capacitors in parallel and in parallel with an additional 10 and 4.7 (Radio Shack No. 272-998) mfd capacitors for a total of 34.7 mfd. The series resonant frequency of the combination comes out to 563.7Hz. An actual measurement showed it to be 520Hz which is within 10 percent, which is the tolerance of the capacitor values.

When the woofer and midrange speakers were connected to the low-pass and high-pass filters just described, the actual response was modified somewhat by the response of the speakers. This effect was particularly noticeable below 50Hz and above 5kHz where the woofer's low frequency response decreased and where the midrange's high frequency response began to fall. See Figure 4-13.

The response of this high-pass filter was measured terminated by an 8 ohm resistor and the results have been plotted in dB on Figure 4-12.

The tweeter was added by merely connecting it through a 3 mfd capacitor across the amplifier output. The responses of the midrange and tweeter speakers provided a virtual cross-over around 4kHz. The response of the complete three way speaker system is shown in Figure 4-13. The resulting relative responses are quite satisfactory except for a 3dB rise in the tweeter response. It was found experimentally that an attenuator consisting of 15 ohms in series and 10 ohms in shunt with the tweeter brought the tweeter response down 3dB as shown by the dotted line in Figure 4-13.

A great deal more could be written about cross-over networks and almost an endless amount of time could be spent in devising band-pass filters, phase equalizing networks and so on, but it is felt that the previous material represents the generally accepted way of handling three way speakers.

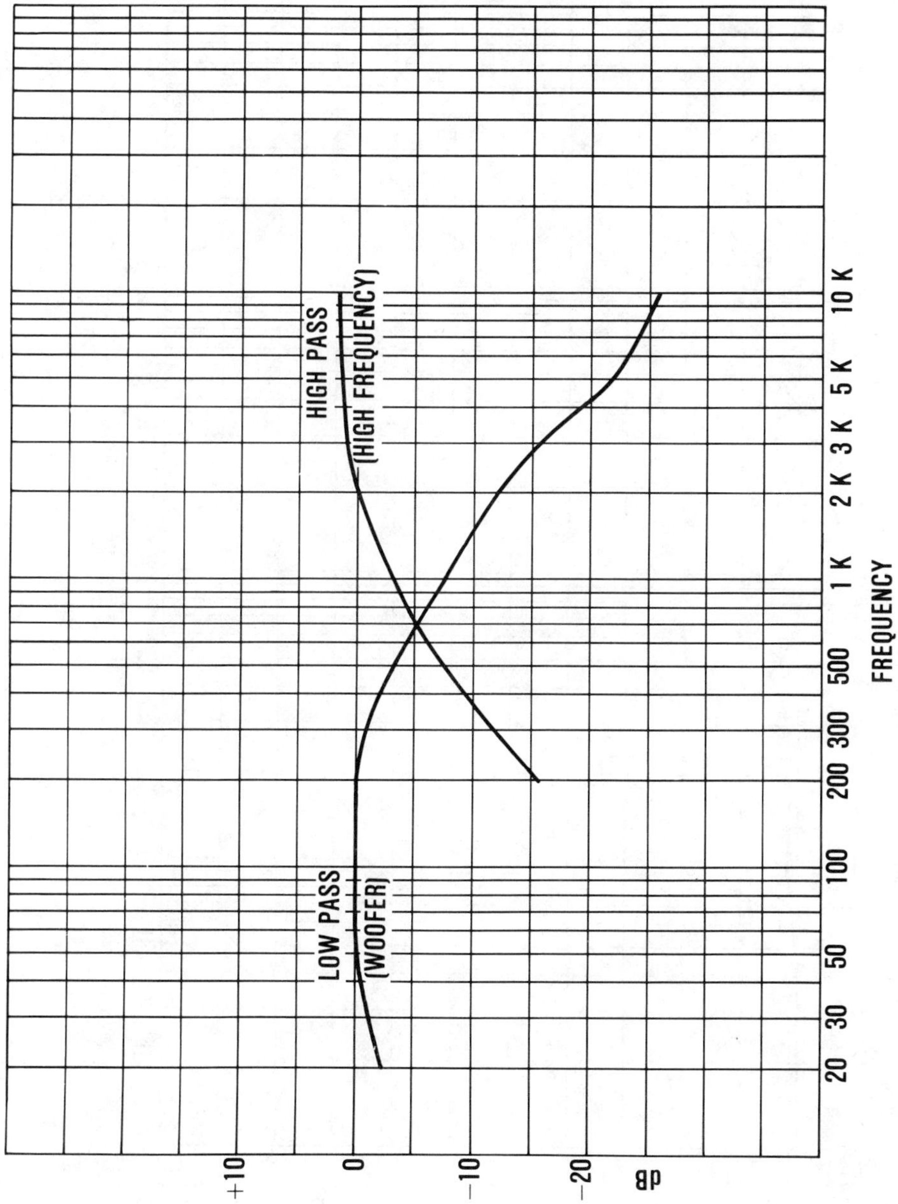
HIGH PASS
(HIGH FREQUENCY)
LOW PASS
(WOOFER)
+10
0
−10
−20
dB
20
30
50
100
200
300
500
1 K
2 K
3 K
5 K
10 K
FREQUENCY

Figure 4-12

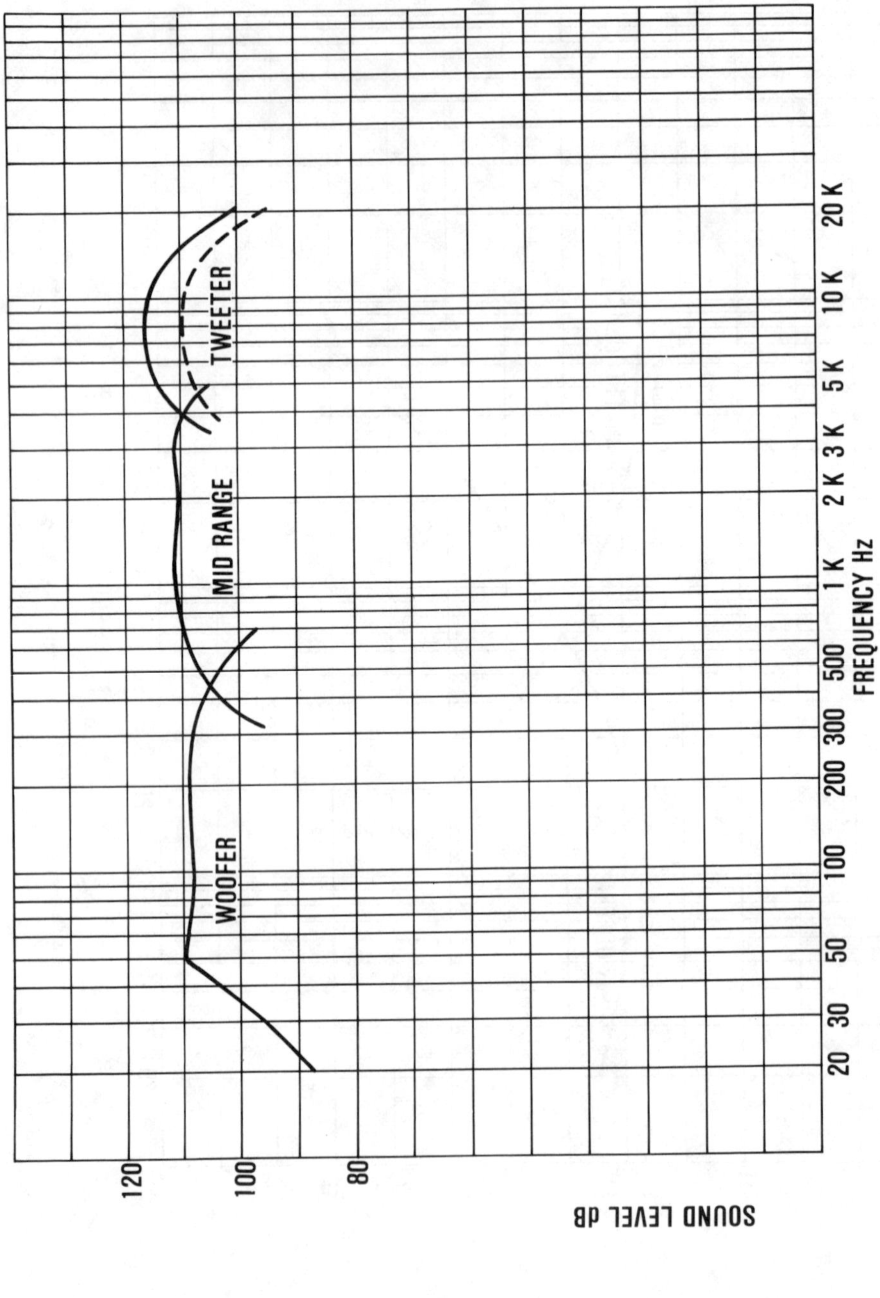
WOOFER
MID RANGE
TWEETER
20 30 50 100 200 300 500 1 K 2 K 3 K 5 K 10 K 20 K
FREQUENCY Hz
120
100
80
SOUND LEVEL dB

Figure 4-13

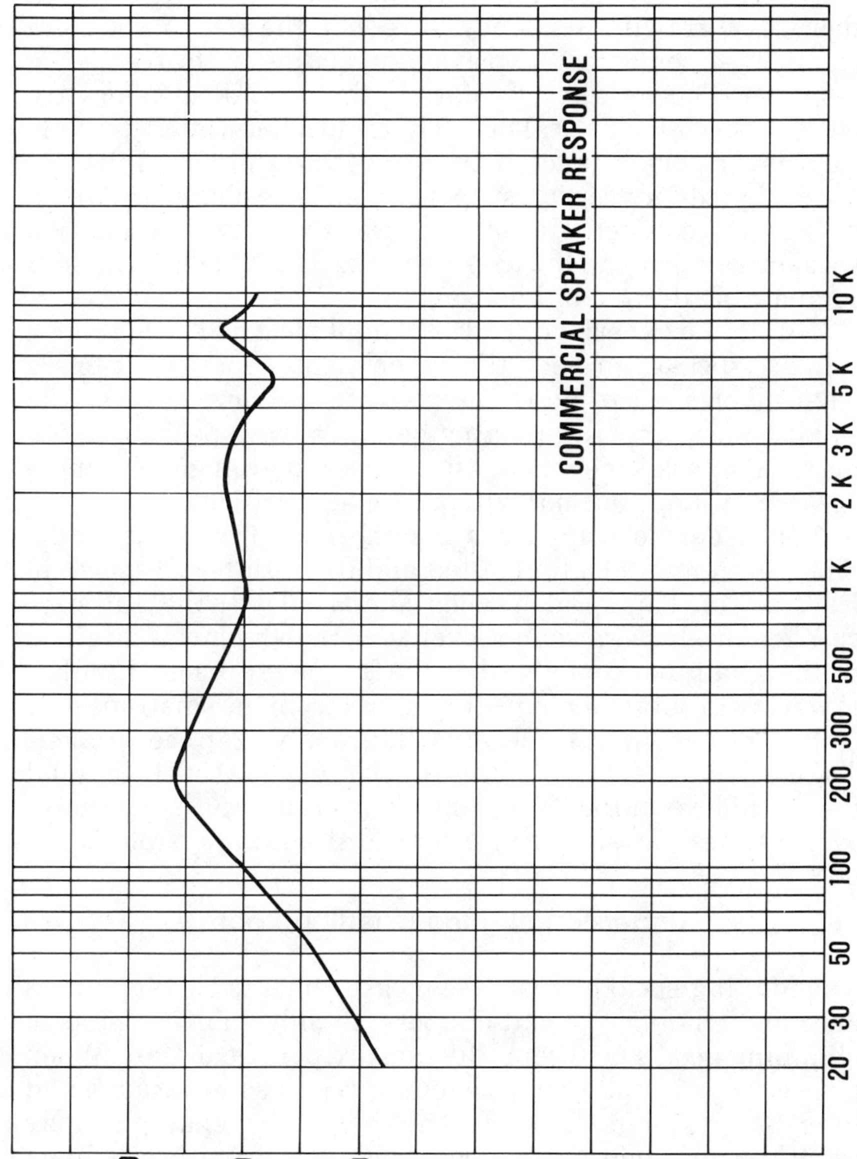

Figure 4-14: Radio Shack Minimus —0.5

Building Smaller Speakers

From observing speakers on the market, it is evident that there is a growing tendency to reduce the size of speakers along with somewhat extravagant claims as to frequency response. The so-called woofers in these small speakers are often 4- or 5-inch speakers. These speakers often provide surprising performance if they have very high compliance, are designed for acoustic suspension and are housed in a firm, well sealed cabinet. However, there is no way that a 5 inch speaker can provide the bass response of a 12 inch speaker of comparable design and housing.

With the experience and techniques learned by making a full size speaker project, the experimenter may want to see what he or she can do in the way of smaller speakers. The smaller speaker will probably be a two way speaker, which makes the task somewhat easier. One can even go to a single speaker system, but not without some sacrifice.

In order to compare a commercial product, a Radio Shack minimus -0.5 was tested and the results are shown in Figure 4-14. This speaker when compared with the "full size" speaker produced a sound level at 45Hz which was attained by the small speaker only when the frequency reached 200Hz. If you try to obtain a room-filling bass from the small speaker by raising the input level, you will probably zap the speaker. The excursion required of the 4-inch speaker at 50Hz to equal the sound level of the 12-inch speaker requires an excursion of over 5 times, physically impractical if not impossible.

Understanding the Listening Room

Making speaker response measurements close up, in an anechoic chamber or out-of-doors, is only the first step in obtaining good sound reproduction in a practical sense. When a speaker is installed in a listening room, some strange and interesting things happen. The room, for example, may greatly strengthen certain frequencies and greatly attenuate others. Echoes and reverberation become very important factors. The acoustic nature of the room and the placement of the speaker or speakers may become a determining factor. Chapter 10 will go into some detail on this subject.

Figure 4-15 is a photograph of the completed Full Size Speaker. Figure 4-16 is a photograph of the speaker and the author standing in front of his laboratory building where all the experiments and construction described in this book were carried out.

Figure 4-15: Full Size Speaker

Figure 4-16: Author with Speaker Outside His Laboratory

5

How to Build Preamplifier, Mixer and Equalizer Projects

Power amplifiers as such, are usually adapted to accept input voltages of 0.5 volt or more. Radio tuners and tape reproducers generally supply such voltages and may be connected directly to power amplifier inputs. However, lower level sources such as phonograph record pickups require additional amplification. Such an amplifier is called a preamplifier.

Why Pre-emphasis is Used in Disk Recording

Phonograph records are made by cutting a master groove with a sharp stylus in a wax or vinyl disk. When such a stylus is fed with constant power, the stylus is moved at constant velocity. This means that as the frequency of the components of music or speech being recorded goes to a lower frequency, the amplitude of the stylus excursion increases proportionally. Thus, the lowest frequency being recorded would have the greatest amplitude. This leads to a very limiting condition where the lowest frequency occupies the full width of a given record groove, and higher frequencies are recorded with decreasing amplitude and quickly become too small to reproduce.

The record industry soon decided that below some predetermined frequency, records should be cut at constant amplitude rather than constant velocity. This preserves practical amplitudes in the sound track cut in the master. The predetermined frequency which was chosen was 300 cycles.

There is another factor which is important and that is noise. This noise is due to slight roughness and irregularities in the record material and tends to increase at higher frequencies. Since the amplitudes of the higher frequency sounds recorded tend to decrease with frequency, the noise problem can largely be overcome by increasing the amplitude of the sounds being recorded as the frequency increases.

The low frequency and high frequency preemphasis have been combined and a standard curve known as the RIAA preemphasis curve has been set up and widely adopted by the disk recording community. This curve is shown in Figure 5-1. Now, in order to restore an overall flat characteristic the playback amplifier must be predistorted with a curve complementing the recording curve. This playback characteristic is also shown in Figure 5-1.

Similarities Between Disk and Magnetic Tape Recording

Recording on magnetic tape is accomplished by passing a tape coated with very fine particles of magnetic material across a gap in a magnetic recording head to which currents are applied representing the sounds to be recorded. This leaves a trail of tiny magnets in the coating, the magnetic strengths of which represent the amplitudes of the sounds being recorded. In other words, the tiny magnets have equal strengths for equal amplitudes of recording current at all frequencies. In order to play back the magnetic recording, a playback head having a gap and a pickup coil is used. The tape is passed across this gap at the same speed with which it was passed across the recording gap. In passing across the gap, the tiny magnets on the surface of the tape generate a voltage in the pickup coil coupled to the gap. But now the

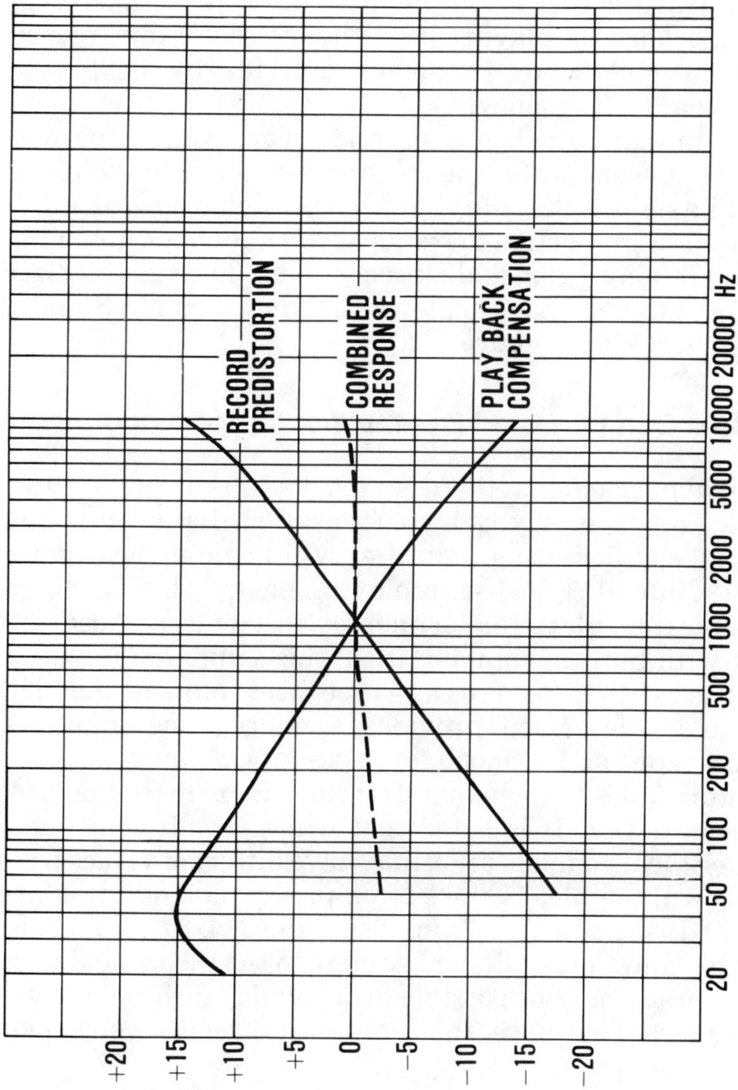

Figure 5-1: Disk Recording Compensations

voltage generated in the pickup coil is not proportional to the strength of the tiny magnets, but is proportional to the rate of change of flux induced in the pickup coil. The magnet representing a low frequency causes a low rate of change compared with the effect of the magnet representing a high frequency. Thus, in playing back the magnetic tape record, the output rises with frequency and ideally is directly proportional to frequency.

To summarize, when magnetic recording is done with constant current into the recording head, it requires preemphasis in the playback in which the output is attenuated in proportion to frequency in a manner much like that used in playing back disk recordings. The standardized curve for magnetic tape playback is known as NAB and is shown generally in Figure 5-2.

Make a Compensated Preamplifier for Tape and Disk

It will be noted that there is a marked similarity between the disk and tape playback compensation. Later on in this chapter, a compensating amplifier will be described that is primarily intended for use in playing back disk recordings. However, although the experimenter will generally use tape players with built-in compensation, this same compensation can be used with little error as a tape head compensator. The mixer to be described provides a microphone input, an auxiliary input and a phono (or tape) input.

Figure 5-3 is the schematic circuit diagram for the three way preamplifier. It is noted that while circuit component values for ideal compensation may have different values from those shown, the ideal values often prove to be nonstandard values and difficult to obtain. The values shown are easily obtainable and the results are so close to the theoretical ideal that it is practically impossible to detect any differences by a listening test. Therefore, the compromise seems well taken.

How to Prevent Low Frequency Disturbances

When using low frequency boost, trouble from low frequency disturbances may be encountered. Perhaps you

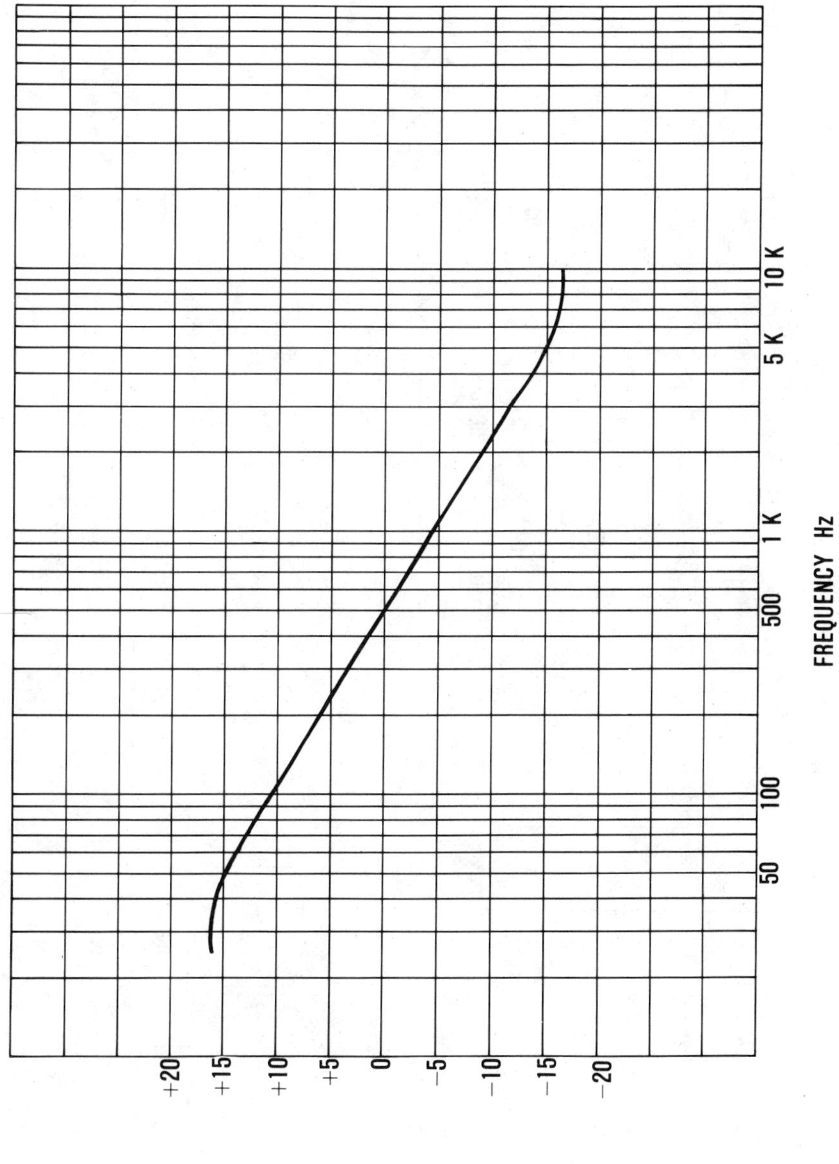

Figure 5-2: Generalized Magnetic Tape Playback Equalization

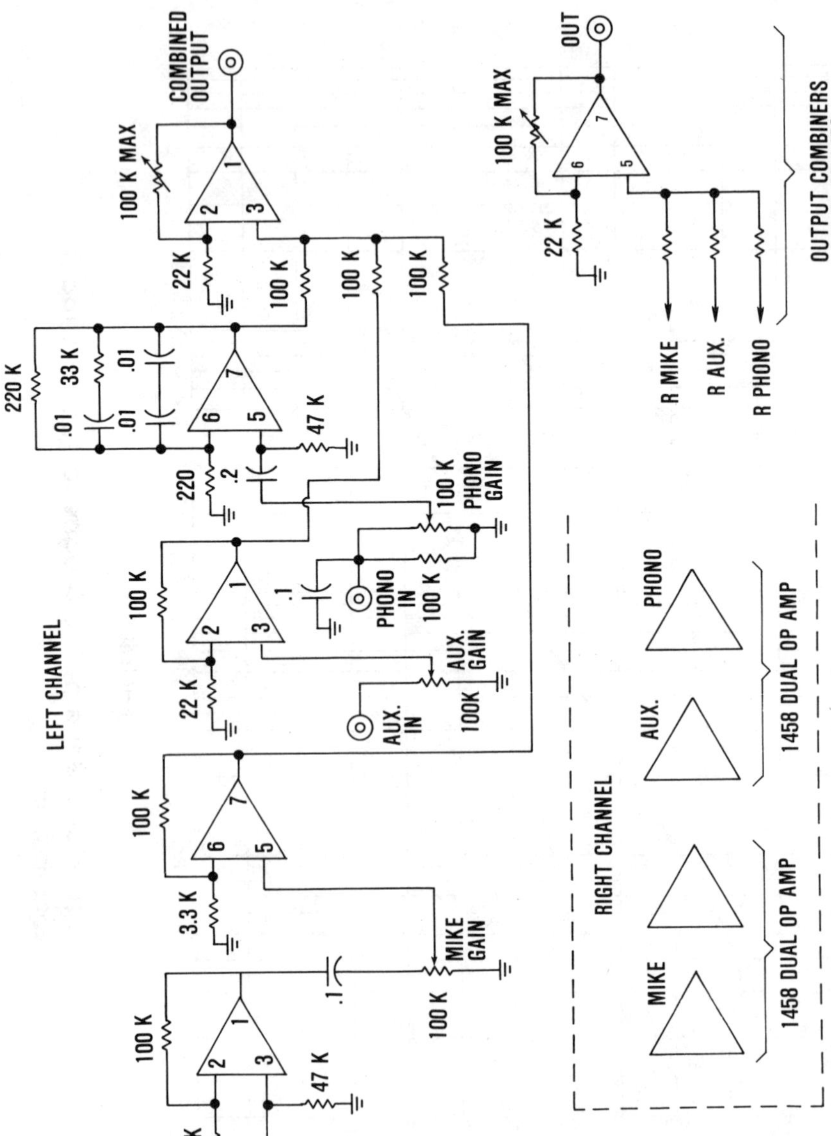
LEFT CHANNEL
MIKE IN
3.3 K
100 K
47 K
.1
100 K
MIKE GAIN
3.3 K
100 K
AUX. IN
100K
AUX. GAIN
22 K
100 K
.1
PHONO IN
100 K
100 K
PHONO GAIN
220
.2
47 K
220 K
.01
33 K
.01
.01
.01
22 K
100 K
100 K
100 K
100 K MAX
COMBINED OUTPUT
RIGHT CHANNEL
MIKE
AUX.
PHONO
1458 DUAL OP AMP
1458 DUAL OP AMP
R MIKE
R AUX.
R PHONO
22 K
100 K MAX
OUT
OUTPUT COMBINERS EACH 1/2 OF 1458 DUAL OP AMP

Figure 5-3

have to compromise between getting the last ounce of low notes and freedom from low frequency disturbances such as turntable rumble. This means there has to be a low frequency roll-off at some point. Such a roll-off is easily and effectively accomplished at the input and output of the preamplifier. Figure 5-3 shows a loading and coupling circuit for the phonograph pickup consisting of a 0.2 microfarad coupling capacitor connected to a 47K shunt resistor. This provides the usual 47K load for the pickup and at the same time provides a low frequency roll-off starting around 20Hz.

Assembling the Compensating Preamplifier

In order to minimize the number of different component parts in the compensating circuits, various combinations of series and parallel capacitors were used. The 0.015 microfarad capacitor was formed of one 0.01 microfarad capacitor paralleled by two 0.01 connected in series. The 0.005 was formed of two 0.01 in series and the 0.0015 was formed of an 0.001 and 0.0005. Thus, the capacitors used were six 0.01 microfarad, one 0.001 and one 0.0005. The resistors were all quarter watt carbon resistors. The complete parts list for the RIAA/NAB compensated preamplifier is as follows:

1—220 ohm quarter watt
1—220K ohm quarter watt
1—3.3K ohm quarter watt
1—33K ohm quarter watt
6—0.01 microfarad ceramic capacitors
1—0.001 microfarad ceramic capacitors
1—0.0005 microfarad ceramic capacitors
1—LM1556V op amp or equivalent

These components were assembled on perf board having 0.1 × 0.1-inch hole spacing. This board accommodates the op amp pins and is otherwise very convenient and easy to use. Colored leads were used for completing the circuit with input/output and other components to complete the circuit as shown in Figure 5-4.

Checking Overall Performance of Disk and Tape Circuits

Figure 5-1 is a series of plots for evaluating the disk compensating preamplifier. One cannot be sure exactly what predistortion characteristics were used in making a given record but the recording industry generally agrees on a curve at least close to the one labeled "RECORD PREDISTORTION." The playback compensated preamplifier described before was measured and the results shown in the curve labeled "PLAYBACK COMPENSATION." When the two curves were added, the result was as shown, dotted and labeled "COMBINED RESPONSE." Thus, if your record has been made according to the accepted convention, the playback with your amplifier (RIAA) compensated for disk will be within ±2dB from 50 Hz to 10kHz which readily meets any practical hi fi criteria.

The situation with the tape playback is not as straightforward. Figure 5-5 shows a curve labeled "IDEAL RECORDING." This presupposes an ideal tape and an infinitely small head gap among other things, neither of which is found in practice. The most significant deviation from the theoretical will be a drooping of the high frequency response which cannot be readily predicted and will vary significantly from situation to situation.

The playback characteristic of the NAB compensation is shown as "NAB COMPENSATION." When this latter is added to the theoretical response, the result is shown in the curve labeled "IDEAL PLAYBACK COMPENSATED."

Providing for Stereo Records and Tapes

Obviously, stereo records or tapes require two of the above described compensating preamplifiers, one for the left channel and one for the right.

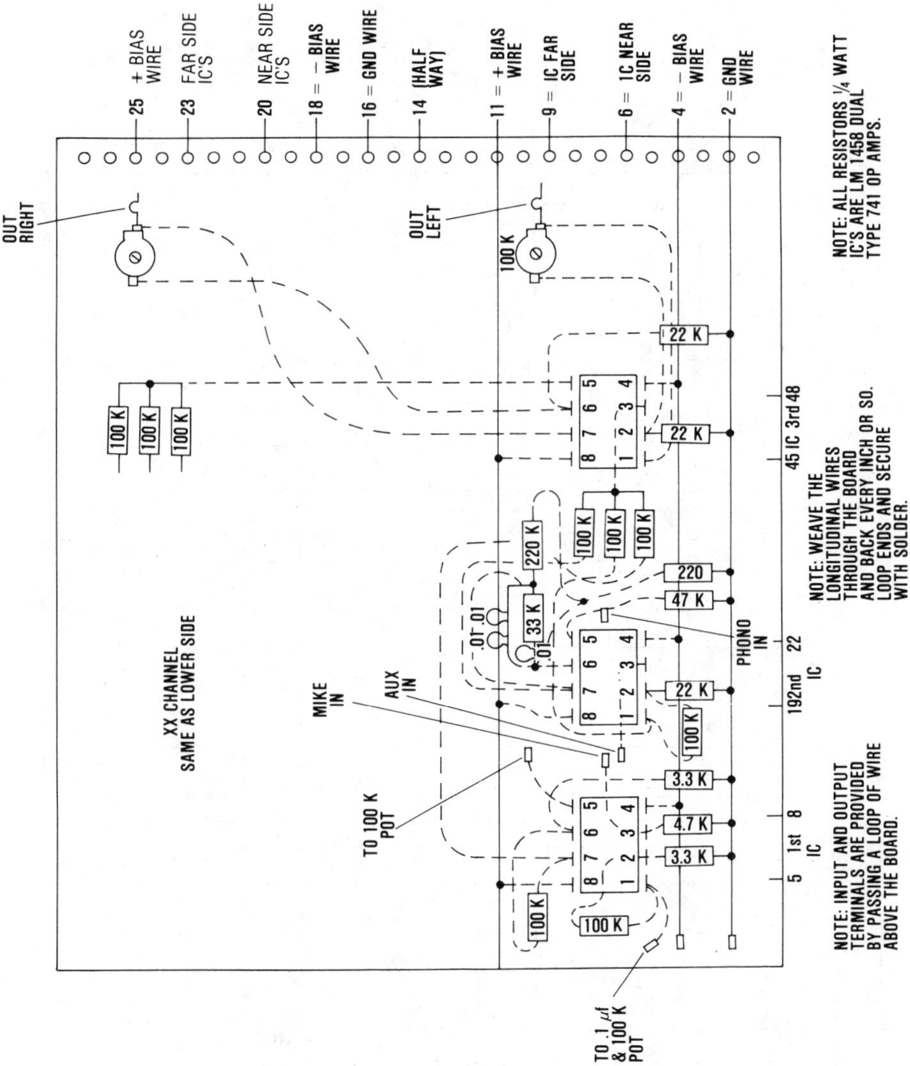
OUT RIGHT
OUT LEFT
25 + BIAS WIRE
23 FAR SIDE IC'S
20 NEAR SIDE IC'S
18 = − BIAS WIRE
16 = GND WIRE
14 (HALF WAY)
11 = + BIAS WIRE
9 = IC FAR SIDE
6 = IC NEAR SIDE
4 = − BIAS WIRE
2 = GND WIRE
XX CHANNEL SAME AS LOWER SIDE
MIKE IN
AUX IN
PHONO IN
TO 100 K POT
TO .1 μf & 100 K POT
5 1st IC 8
19 2nd IC 22
45 IC 3rd 48
NOTE: ALL RESISTORS ¼ WATT IC'S ARE LM 1458 DUAL TYPE 741 OP AMPS.
NOTE: WEAVE THE LONGITUDINAL WIRES THROUGH THE BOARD AND BACK EVERY INCH OR SO. LOOP ENDS AND SECURE WITH SOLDER.
NOTE: INPUT AND OUTPUT TERMINALS ARE PROVIDED BY PASSING A LOOP OF WIRE ABOVE THE BOARD.

Figure 5-4

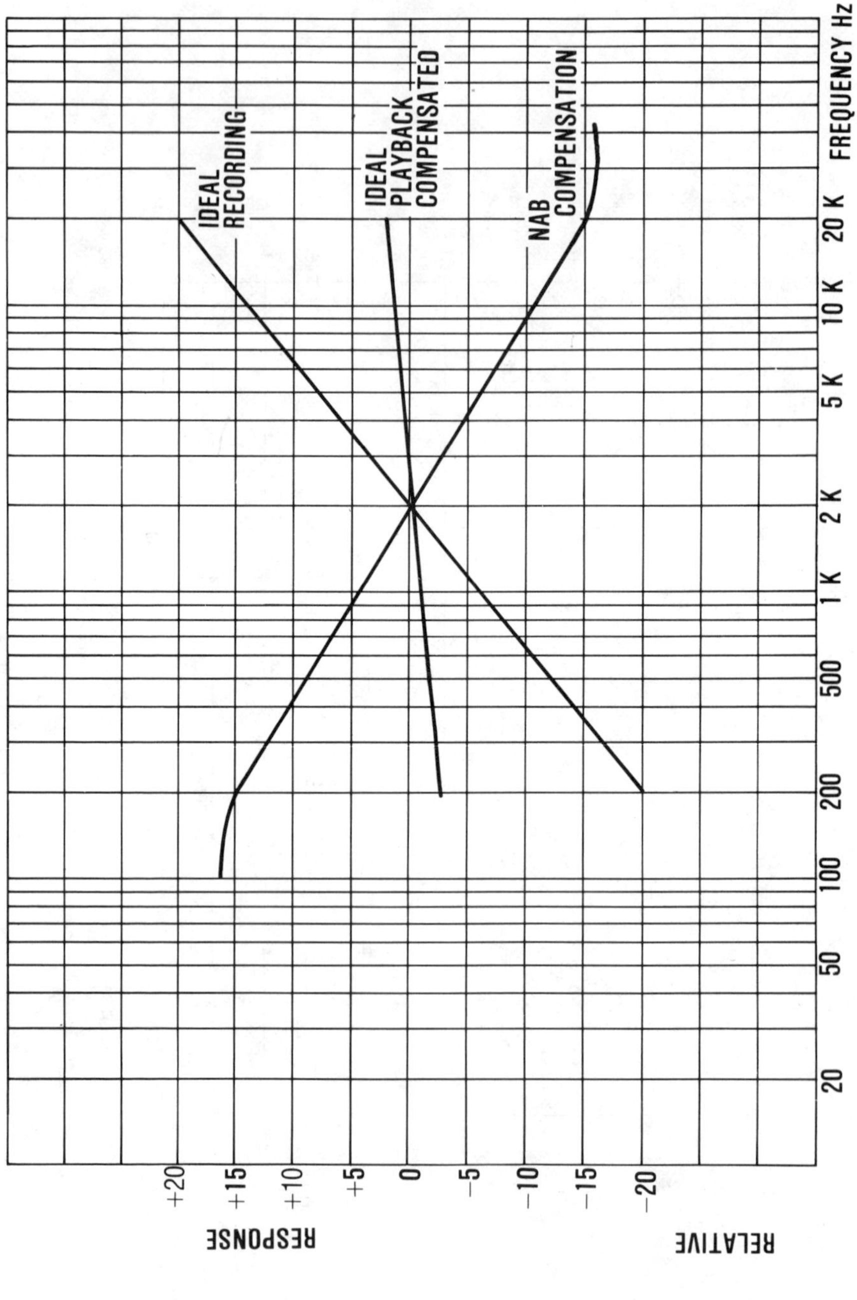

Figure 5-5: Magnetic Recording Compensation

Mixing Sound Signals from Various Sources

Many interesting experiments may be carried out with a mixing panel. A sing-along with tape or phonograph is a typical example. Mixed sounds may be reproduced over a loudspeaker or recorded on tape. Announcements may be added to a tape recording or in rerecording from disk, and editing of a tape or disk can be carried out.

Versatile Mixing of Mike, Phono and Auxiliary Sounds

Mixing sound signals from several sources can be both interesting and useful. Let us consider the combination of microphone, auxiliary sources and phonograph inputs. The auxiliary input adds more than one dimension to the combination since it may be derived from a tape player, radio tuner or other source of moderate level. When all of these are stereo sources, still further combinations are possible.

Sing Along with Stereo Phonograph or Tape Recorder

With stereo microphones (including a left channel microphone and a right channel microphone) and phonograph, you can sing along with the left channel, or two people can sing along, one on the left and the other on the right. This can be done for entertainment over loudspeakers or the results can be recorded on a stereo tape recorder for future enjoyment.

Editing a Magnetic Tape

Often times a magnetic tape can be considerably improved by editing. In some cases, you can add interest by interspersing announcements of what is to follow. The above can be accomplished by playing the tape to be edited into the auxiliary input and tape recording the output. Editing may consist of omitting some passages or rearranging others. If the player is stopped at a given point, an announcement may be recorded over one of the microphones and the playing then continued.

Sing Along with the Radio

If a radio tuner is connected to the auxiliary input, you can sing along with the radio. Again, this can be played over loudspeakers or recorded on tape.

Brief Description of the Mike, Phono, Tape Equalizer/Mixer

This versatile combination is quite readily provided, including left and right stereo channels. First, the two microphone channels are provided for by means of two dual 741 (1458) type op amps. By operating the two halves of the op amp in cascade at moderate gain, sufficient overall gain is achieved with a reasonable bandwidth, about 20kHz. A second dual 741 (1458) for each channel provides a low gain auxiliary amplifier and an RIAA compensated phonograph amplifier. The three amplified sources are combined in a fifth dual 741 (1458) which sums the left channel sources in one half and the right channel sources in the other half. A separate gain control is provided for each mike, a stereo gain control for the auxiliary inputs and a stereo gain control for the phono inputs. The combining amplifiers are provided with internally adjustable gain controls for equalizing overall gain in the two channels.

Figure 5-6 is a drawing of the front panel of the mixer amplifier showing gain controls and input jacks for each function for a mono system. A separate gain control should be provided for the left and right microphones. Stereo gain controls should be provided for the tape/tuner/auxiliary and phono inputs. Balancing can be carried out using the two gain controls on the main amplifier panel.

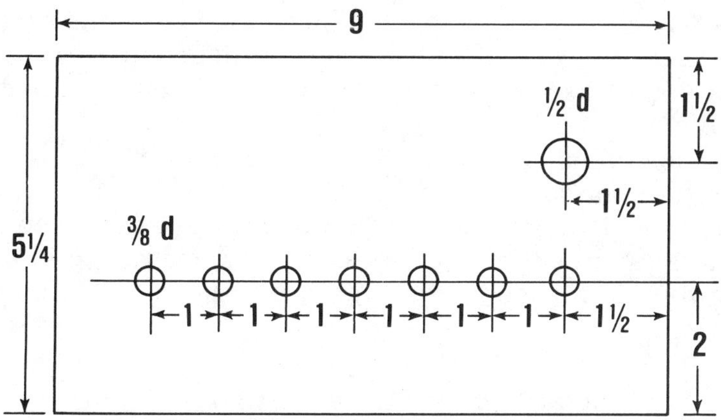

Figure 5-6

Details of the Mixer Circuit

Figure 5-3 is a detailed schematic of the mixer circuit for one channel. The second channel is identical. Five (LM1458) dual 741 op amps were used because this amplifier is one of the most readily available. When carefully used, it provides quite satisfactory results. All stages except the mike stages are low gain and, hence, have a wide frequency response. For example, the auxiliary amplifier with a gain of 15dB is down 3dB at 90kHz.

The first dual op amp in each channel is used as the microphone amplifier. By limiting the gain in each of two stages to 32dB (40X) and cascading two stages, an overall gain of 64dB (1600X) is obtained and a response down 3dB at 20kHz. The R-C coupling between the stages cuts infrasonic signals and provides for a gain control.

The second dual op amp provides the tuner/tape/auxiliary buffer amplifier and the RIAA compensated phono amplifier, one dual amplifier in each channel. The gain controls for each are stereo (dual) controls since it is assumed that with stereo inputs, both gain controls should be advanced together. Stereo balancing can be carried out with the main amplifier gain controls. The phono input is capacity coupled in order to reduce infrasonic signals such as turntable rumble.

The fifth dual op amp is used as a pair of combining amplifiers for combining the various signals and outputting them. The two halves of this op amp are provided with adjustable gain controls for setting the overall gain of the system and for equalizing gain in the two channels if required.

Assembling and Wiring the Mixer Board

The two-channel mixer described above was assembled on a $2\frac{3}{4} \times 6$ inch perf board with 0.1X0.1 hole spacings. This board had 27 rows of holes with 57 holes in each row. Figure 5-4 is a drawing of the assembly and wiring of the mixer. Numbers along the borders show the hole numbers of the rows and columns where wires and parts are placed. The view is from the top of the board and wires along the underside are shown as dotted lines. Notes indicate input and output terminals and provide other instructions for fabrication. The op amps (ICs) are assembled by pushing their pins through holes as indicated by the horizontal and vertical hole numbers. When the pins have been pushed through, bend them away from the body of the IC and tin for easy and sure soldering. A small chisel pointed soldering iron should be used and care should be taken that solder does not bridge from one pin to the next. After soldering all connections, inspect with a microscope and check carefully against the circuit diagram.

Testing the Mixer

Before assembling the mixer in a cabinet, it would be well to examine its overall performance. This testing should include sensitivity, frequency response and hum measurements, although the latter will be modified when the amplifier is mounted in a metal cabinet. The model described had gains at 1kHz of 60dB at the mike input, 17dB at the auxiliary input and 40dB at the phono input. The frequency characteristic of the mike amplifier was as flat as could be measured to over 10kHz and down 1dB at 20kHz. The auxiliary input was flat to 1kHz and showed a gradual rise to plus 1dB at 20kHz. The phono input was very close to the curve taken on the circuit before and plotted in Figure 5-1.

Assembling Mixer in Metal Cabinet

The mixer described was mounted in a Radio Shack Cat. No. 270-0270 metal cabinet $9\frac{1}{4} \times 5\frac{3}{8} \times 6\frac{3}{4}$ inches. The first unit built was a mono mixer as shown in Figure 5-7. The inside assembly view is shown in Figure 5-8. The front panel drilling is shown in Figure 5-6. The placement of the input and output jacks between the gain controls proved rather inconvenient and so the experimenter is advised to work out a better arrangement particularly for his or her stereo mixer.

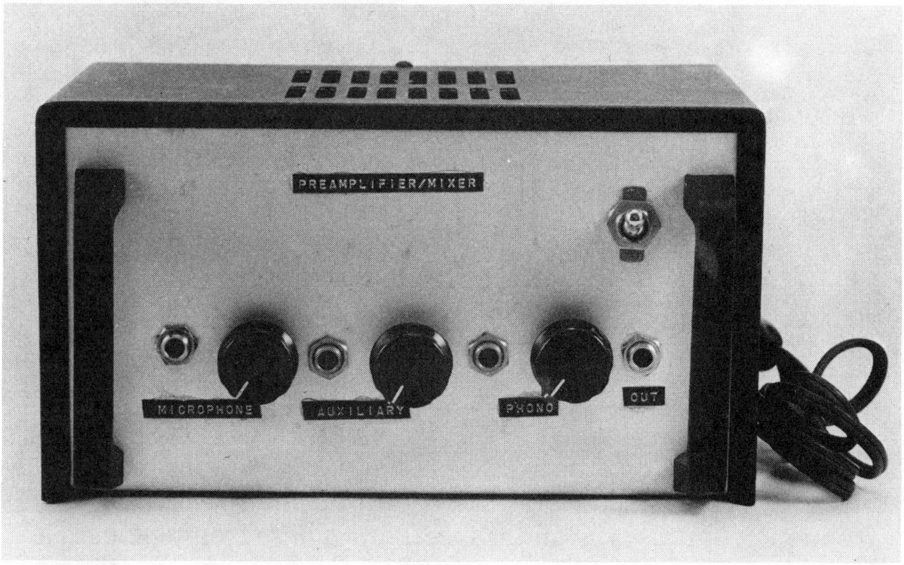

Figure 5-7: Mixer, Mono Version

Figure 5-8: Mixer, Interior

Build a Five Channel Equalizer

An interesting and most useful device is a multiple band equalizer. When slide type controls are used they can simulate the gain or attenuation characteristics of the equalizer and are called "graphic equalizers."

Equalizers of these types provide resonant gain or attenuation characteristics at a number of predetermined frequencies across the audio spectrum. These frequencies are chosen and spaced for use in emphasizing or deemphasizing preselected bands of frequencies from some low frequency near the lower end of the spectrum to some high frequency near the upper end of the spectrum. Typically, equalizers provide a maximum peak or attenuation of 12dB with smooth adjustment through the range.

Equalizers have a number of practical and interesting uses. They can be used to correct deficiencies in the response of various parts of a system such as tape recordings and loudspeaker response. They can be used to color sound reproduction, making it more brilliant and/or more boomy. Room acoustics can often be improved with an equalizer. Prerecorded material can be doctored as by decreasing hollow sounding reproduction caused by reverberation in the recording room. Copies on tape can be improved by countering the frequency limitations of the original which is often compounded in the copy.

Figure 5-9 shows a basic equalizer circuit together with a table of L and C values for boost/attenuate frequencies of 60, 240, 1000, 3500 and 10,000 Hz. The proposed boost peak and attenuation maximum at the resonant frequencies is 12dB (4 times). This boost and attenuation maximum will depend on the effective Q of the series circuit. Resistor R can thus be adjusted or chosen to provide a Q which provides the desired boost/attenuation. With R determined using a variable resistor, an equivalent fixed resistor may be used in the final assembly.

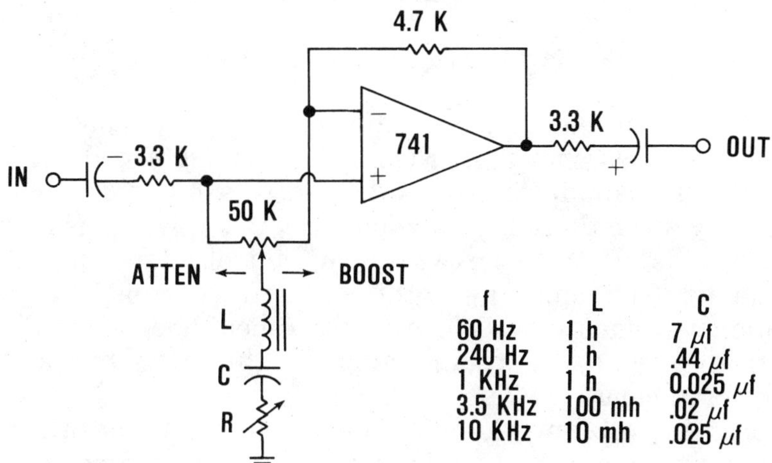

f	L	C
60 Hz	1 h	7 μf
240 Hz	1 h	.44 μf
1 KHz	1 h	0.025 μf
3.5 KHz	100 mh	.02 μf
10 KHz	10 mh	.025 μf

Figure 5-9

Figure 5-10 shows the complete circuit for a five equalizer assembly. A second identical equalizer completes a stereo system.

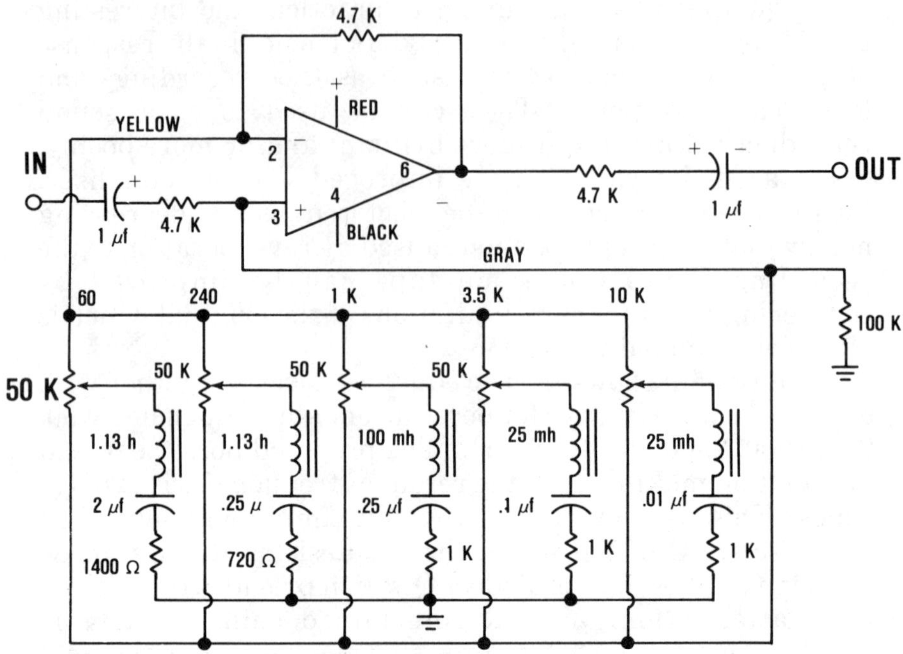

Figure 5-10

Putting the Equalizer Together

Since a stereo equalizer requires two sets of controls, a panel layout as shown in Figure 5-11 is suggested. Some equalizers provide side-by-side controls but this seems confusing since there is no obvious way to tell which is right and which is left. With the two rows of controls, it seems easier to identify not only the right and left controls but the frequency ranges as well. The controls should be mounted on the front panel as shown and labeled right-left and with the peak frequencies.

Some problems may be encountered in obtaining suitable inductors. For the large inductor, the primary of a Radio Shack output transformer Cat. No. 273-1380 was tried.

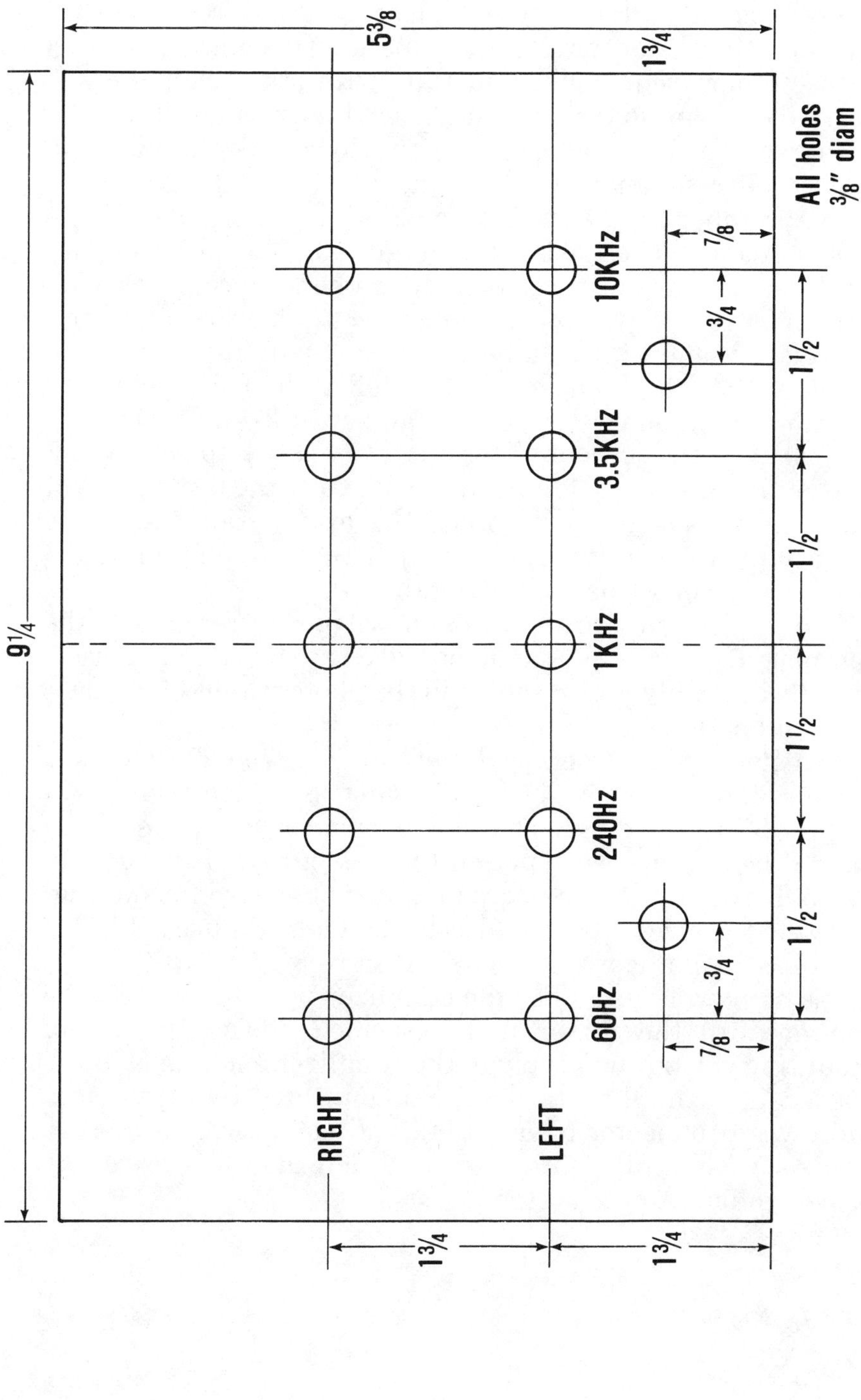

Figure 5-11: Equalizer Front Panel Drilling

It resonated with 2 microfarad at 70Hz and with 0.25 microfarad at 270Hz, near enough the selected frequencies of 60 and 240Hz for practical purposes. Of course, one can choose other frequencies and tune with other capacitors if desired. The audio transformer secondary was tested for the higher frequencies but showed multiple peaks, apparently due to self-resonance of the primary.

A number of other coils were tried. For example a Radio Shack Xeon strobe flash tube trigger coil looked promising but turned out to be unsatisfactory. A 1 henry toroid was tried but the response was too sharp and could not be broadened sufficiently and at the same time provide the desired peak boost. A 100 mh r.f. choke worked well at the 1kHz point and two 25 mh r.f. chokes were satisfactory at 3.5 and 10kHz.

Each of the five equalizer circuits was set up and tested. The results are plotted in Figure 5-12. Only the boost parts of the responses are shown since the attenuation parts are essentially mirror images in which the attenuation in dB matches the boost parts in dB gain.

A perf board was drilled as shown in Figure 5-13 and the components were assembled and wired as shown in Figure 5-14. The assembly is also shown in the photographs of Figures 5-15 and 5-16.

A Radio Shack metal cabinet Cat. No. 270-0270 was used to house the equalizer. The front panel was drilled as shown in Figure 5-11. The single channel circuit board was mounted inside the cabinet as shown in the photograph Figure 5-15. The 50K potentiometers were mounted on the front panel and wired to the circuit board, also as shown in Figure 5-16. The completed equalizer is shown in Figure 5-17.

The power supply for the equalizer was the same as the one used in the mixer and described above. Instead of mounting the power supply in the equalizer enclosure, it was housed in a small metal box and connected by means of a three-wire connector (stereo plug and jack, for example).

As with similar projects, a second identical assembly provides for a stereo system.

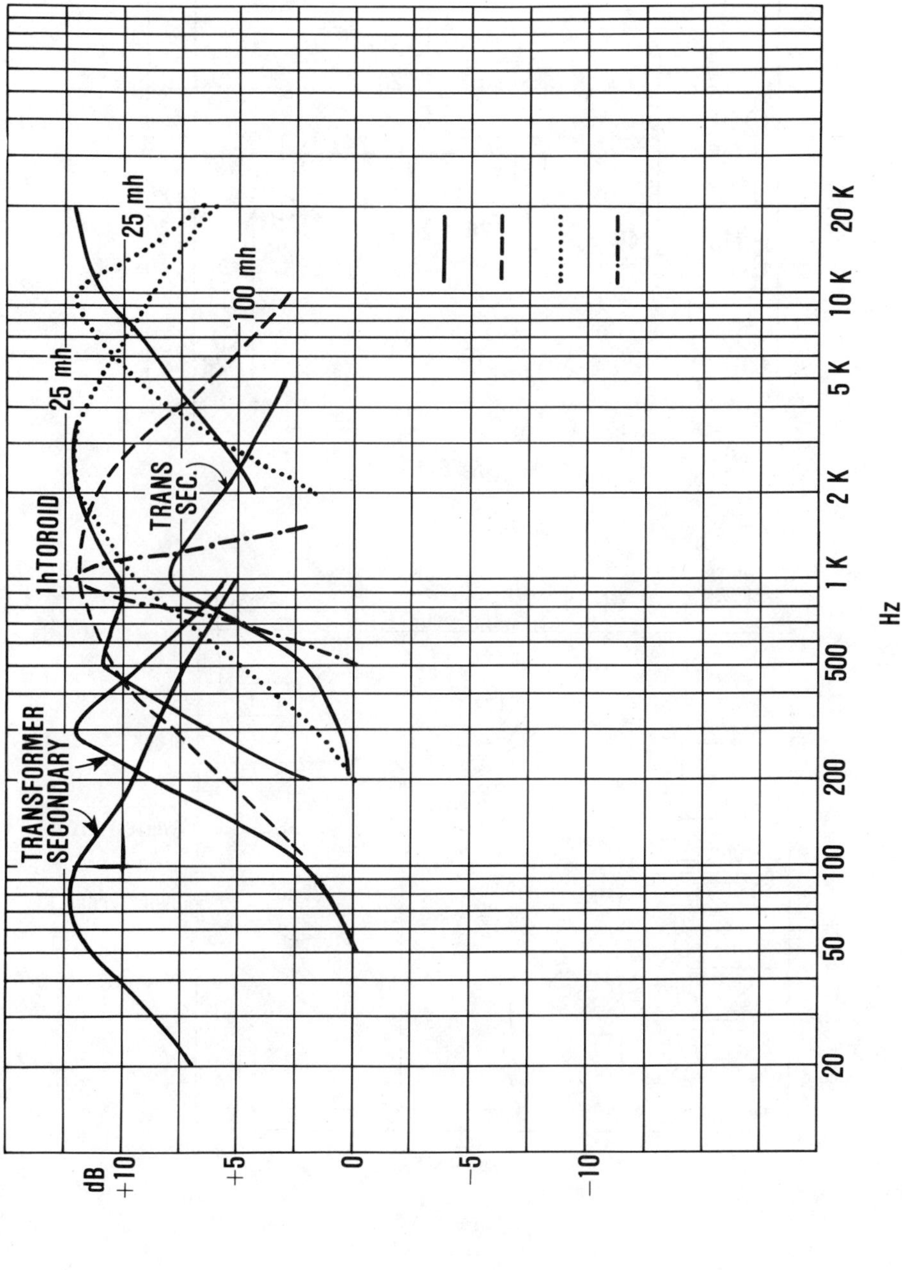
dB
+10
+5
0
-5
-10
20
50
100
200
500
1 K
2 K
5 K
10 K
20 K
Hz
TRANSFORMER SECONDARY
1hTOROID
25 mh
25 mh
100 mh
TRANS SEC.

Figure 5-12

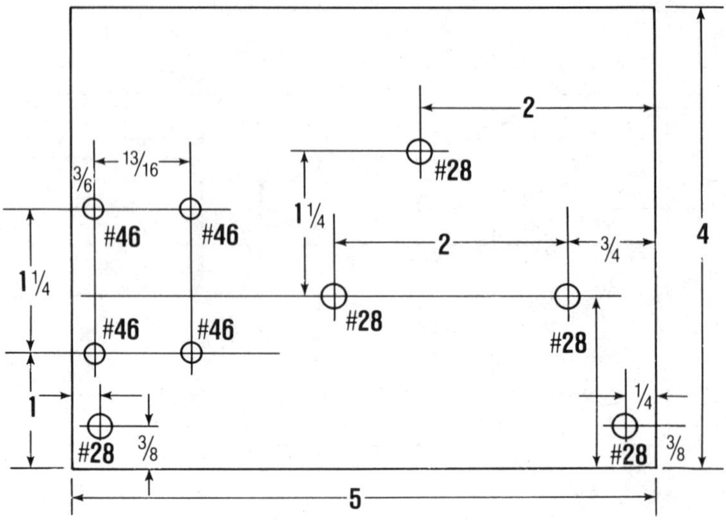

Figure 5-13

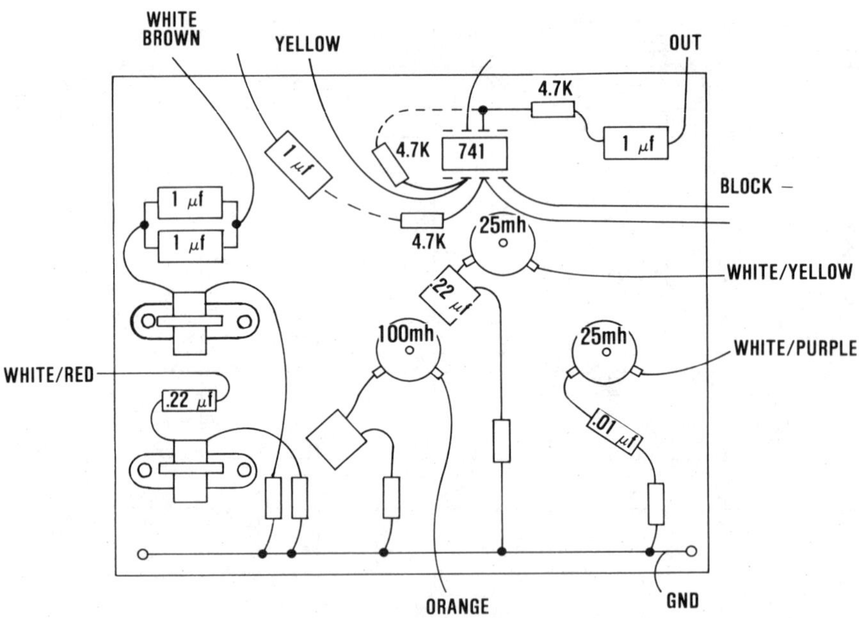

Figure 5-14

Figure 5-15: Equalizer Circuit Board

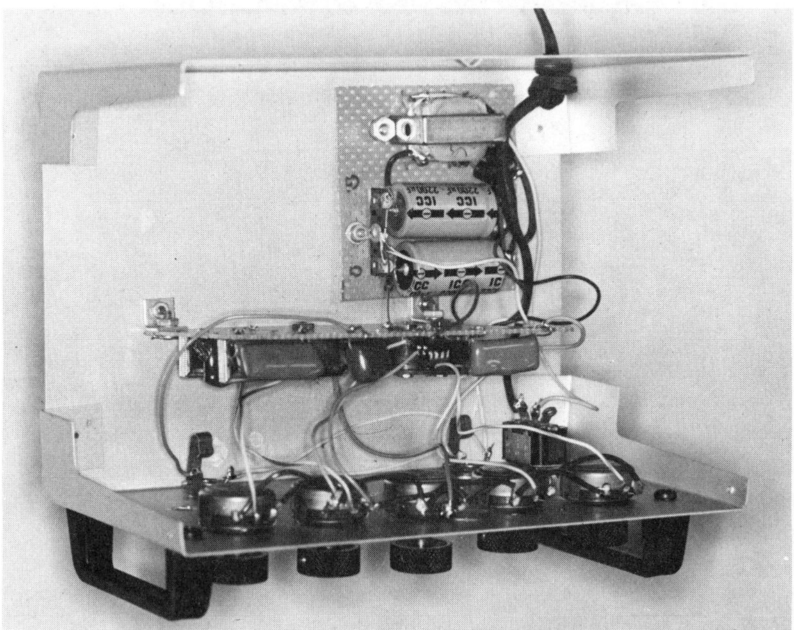

Figure 5-16: Equalizer, Interior

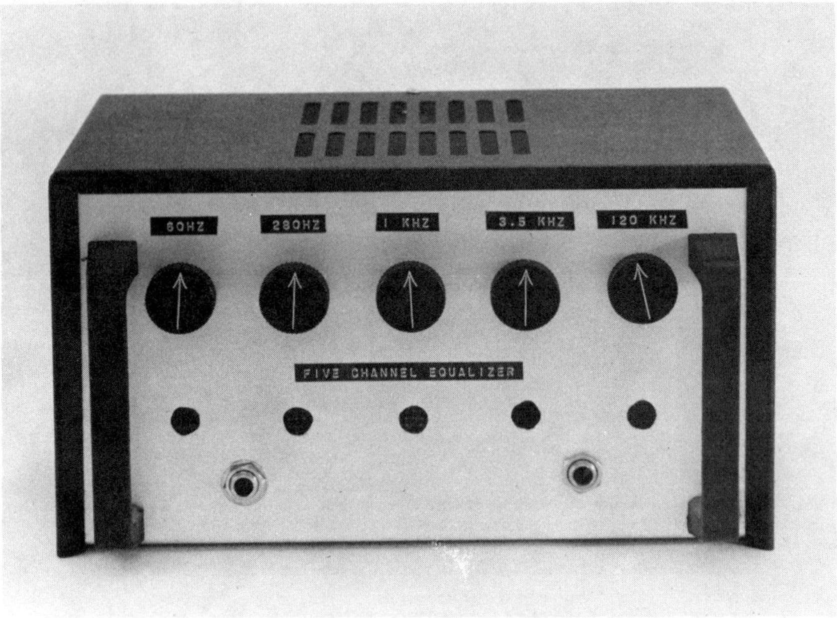

Figure 5-17: Five Frequency Equalizer

6

Building Test Equipment Projects

The Significance of Distortion

Ideal amplifiers, recorders and reproducers produce no non-linear distortion. That is, the output of each or a combination of components is strictly proportional to their inputs. However, there are no ideal components or systems since all produce more or less distortion, which is, non-linear relationship between input and output. In some components, notably high grade amplifiers, the distortion can be reduced to levels too low to detect by the human ear. In some cases, again notably in amplifiers, non-linearity can be measured down to and below the level which can be detected by the human ear.

The two most significant and the most often measured and specified kinds of distortion are harmonic distortion and intermodulation distortion. Harmonic distortion is caused by nonlinearity in some part or most parts of a reproducing system. Harmonic distortion of a pure sine wave or tone results in harmonics, that is, waves or tones having frequencies which are two, three or more times the frequency of the original tone. The audible effect is to color the tone; in the case of music, for example, to make it sound more shrill.

The amount of harmonic which can be detected by ear depends on the order of the harmonic, third being more easily detected than second and so on. The percentage which can be detected by ear is rather small for a pure tone but becomes substantially larger for complex tones, which are encountered in most music. It ranges from a fraction of a percent for pure tones to several percent for complex music.

Characteristically, harmonic distortion may be higher at very low levels in an amplifier, and decrease and stay very low as the level is increased until a high level point is reached at which the distortion starts to increase rapidly to become intolerable as the amplifier is overloaded. Thus, the three significant levels are very low, midrange and very high or overload. Amplifier specifications generally are specified for harmonic distortion at their rated maximum output, that is, the highest level attainable before overload sets in. Unfortunately, the very low level distortion may be unacceptable and is generally unspecified unless a complete characteristic is shown in graphical form.

Again, most often a rating is given in terms of THD (total harmonic distortion). This does not show, for example, that the dominant harmonic may be the third, and hence, more disturbing than the second. On the other hand, harmonic distortion in high quality amplifiers is generally so low as to be completely negligible. This is not necessarily true for tape recording or loudspeakers.

There is a practical problem in measuring harmonic distortion, particularly at very low percentages. The problem is that such measurements require a very pure source of sine wave signals. Providing such a source may prove difficult and expensive.

The second significant method of measuring distortion is to measure intermodulation distortion. It is more easily accomplished with much less expensive equipment. Although there is no direct correlation between the values of harmonic distortion measured on a given component or system and its intermodulation distortion, either one may be used as a valid rating means. Since intermodulation distortion produces discordant components, it is more disturbing than harmonic distortion and may thus be a more valid measure of audio quality.

Intermodulation distortion is modulation of one tone by the other. It is the characteristic that generates sum and difference frequencies when two tones are passed simultaneously through a component or system. The easiest way to measure intermodulation distortion is to use a high frequency tone and a low frequency tone. The two tones are provided in a predetermined amplitude ratio and passed through the device being tested. The low frequency tone is eliminated by means of a high-pass filter which passes only the modulated high frequency. This latter is detected or rectified to derive the low frequency modulation and the remaining high frequency component is eliminated by passing the rectified signal through a low-pass filter. The remaining low frequency is measured. With proper calibration, this measured low frequency can be expressed in terms of percentage of intermodulation distortion. While harmonic distortion and intermodulation distortion cannot be readily quantitatively related, the same non-linear characteristics of a system are responsible for both. Thus, in a way, the measurement of one is as indicative of a system's linearity as the other. The advantage of making intermodulation measurements is that they are easily carried out and with relatively inexpensive equipment.

Measuring Distortion

There are many ways in which distortion of audio signals can be measured and evaluated. All components of an audio system are subject to such evaluation. Some may appear to be nearly flawless in one area only to fail miserably in another. There is no one completely definitive or overall test which will insure truly high fidelity reproduction. Many claim that the only meaningful test is a listening test. Then we must ask, who is the perfect listener? One fallacy of the listening test is that it can seldom pinpoint the source of inaccuracy in reproduction. Without knowing this, how can one take corrective measures?

It seems reasonable to start with electrical tests and to carry electrical quality as far as possible before resorting to listening tests. There are a number of accepted electrical tests to begin with. The more commonly used are as follows:

Frequency response
Power output
Harmonic distortion
Intermodulation distortion
Transient distortion
Phase distortion
Hum and noise
Damping factor

There are further characteristics which are important where electromechanical devices are involved such as loudspeakers, phono pickups, turntables, tapes, tape recorder/players and microphones.

High fidelity frequency response, i.e., 20Hz to 20,000Hz flat within ±2dB is easily attained and can generally be taken for granted in any reasonably good amplifier. Power output can be easily measured but is not a "magic number" due to several contributing factors. One that strikes home immediately is loudspeaker efficiency. See Chapter 4 for more on this subject.

Hum and noise figures (the hum and noise which the amplifier adds to the desired signal) are readily obtained.

Damping factor is the ratio of the load impedance to the internal impedance of a power amplifier.

There remain the four types of distortion. The most commonly specified is harmonic distortion called THD or total harmonic distortion. This is usually measured by suppressing a fundamental frequency test tone and measuring what remains. There are two problems here: One is that such measurements do not distinguish between the various orders of harmonic. In the typical high quality push-pull output amplifier, the major harmonic is generally the third, with the second being suppressed by the balanced output circuit. The higher the order of the harmonic, the more disturbing it becomes. The second serious practical problem is that harmonic distortion measurements require a signal source which is substantially free of harmonics. Such signal sources are not easy to come by and may be expensive.

Phase distortion, also, requires special and expensive equipment. In addition, there is no general agreement on what is tolerable. More often phase distortion is thought to be rather unimportant or at least not troublesome in a good amplifier.

Transient intermodulation distortion is a little known factor. Two easily made tests yield some qualitative information. One is the square wave test, described elsewhere in this chapter. The other is slew rate, which is a major factor in transient intermodulation distortion. The higher the slew rate expressed in volts per microsecond, the less the possibility of its effect on TIM.

The final distortion to be considered is intermodulation distortion. Such distortion concerns the modulation of one component tone by another tone. It is similar to the well known modulation of a radio frequency carrier by an audio signal. It produces sidebands, i.e., sum and difference frequencies. The reason it is so significant in the reproduction of music, is that these sum and difference frequencies sometimes produce discordant tones. While simple harmonic distortion may only affect the quality of individual tones, intermodulation distortion produces new tones which are not only undesirable but more readily detected.

Intermodulation distortion is measured by measuring the modulation of a relatively high frequency tone by a relatively low frequency tone. The reasoning behind this is that intermodulation distortion is easily measured this way and that such measurements give a good indication of the overall performance of an amplifier. One of the related instruments described in this chapter is an Intermodulation Meter.

Build an Intermodulation Meter

As has been stated, an Intermodulation Meter is a useful instrument for measuring low levels of distortion in a high fidelity system. In my opinion, intermodulation distortion is the most meaningful factor in terms of degradation in a high fidelity system. If intermodulation distortion is low, other forms of distortion are almost certain to be low as well. Fortunately for the hi fi experimenter, an Intermodulation Distortion Meter, while being a challenge, is not too difficult to build.

Outlining the Basic IM Meter Circuit

The source signal for the meter is composed of a high frequency signal and an added low frequency signal. The high frequency signal is 5000Hz generated by a Wein bridge oscillator. The low frequency signal is 60Hz derived from the power transformer. Controls are provided for adjusting the relative amplitudes of the two signals. These two signals are mixed in a unity gain amplifier. The output is applied to a step attenuator, the voltage across which may be measured on the indicating meter. This permits setting the output to a value compatible to the equipment undergoing the test.

The equipment being tested thus receives a high frequency signal and a low frequency signal which are added, but not intermodulated. A system without intermodulation distortion characteristics passes on these two signals without change except in amplitude. However, if the system distorts, the low frequency signal modulates the high frequency signal, producing a typical modulated signal, that is, the high frequency signal acquires side bands of + and - the low frequency signal. A measurement of the effective modulation is an indication of the degree of intermodulation distortion produced by the equipment being tested.

The output of the equipment being tested is applied to a potentiometer which is used to adjust the level of signal applied to the intermodulation meter. The percent of IM distortion is measured by measuring the low frequency modulation of a predetermined signal amplitude.

The signal from the equipment being tested is first passed through a high-pass filter. This passes only the high frequency modulated signal and eliminates the low frequency signal. The modulated high frequency signal at this point is adjusted to a predetermined level by means of the input potentiometer and applied to a demodulator. This demodulated signal now comprises the high frequency carrier and the demodulated low frequency. This signal is passed through a low-pass filter eliminating the carrier and leaving only the low frequency product of intermodulation distortion. This latter is measured on the AC voltmeter. By knowing the percentage modulation of the signal applied to the equipment under test, the amplitude of the modulated signal at the demodulator and the amplitude of the demodulation product low frequency signal, you can calculate the percentage of intermodulation distortion.

The percentage modulation of the high frequency signal by the low frequency signal is their ratio in percentage. For example, a ratio of 3 to 1 is 33 percent modulation.

The amplitude of the modulated signal at the demodulator is measured on the AC meter.

The amplitude of the demodulation product low frequency is the low frequency output measured on the AC meter multiplied by the output attenuator ratio and divided by the low frequency gain between the demodulator and the output attenuator.

Intermodulation Meter Circuit Details

Figure 6-1 is a simplified block diagram of the IM meter and Figure 6-2 is a detailed schematic circuit diagram of the completed Intermodulation Distortion Meter. The Wein bridge oscillator employs one-half of a 1458 op amp. Feedback, and hence, the point of oscillation is controlled by a 1K rheostat mounted on the front pane. The oscillation level is stabilized by a FET connected as an automatic level control. The oscillator output is rectified and the resulting DC is filtered and applied to the FET gate. The drain to source impedance is connected in series with a 200 ohm fixed resistor thereby controlling the input resistance of the op amp and hence its gain. Any tendency for the oscillator level to increase, increases the bias on the FET and thereby its impedance, reducing the op amp gain and the oscillation level. The Wein bridge capacitors and resistors are calculated

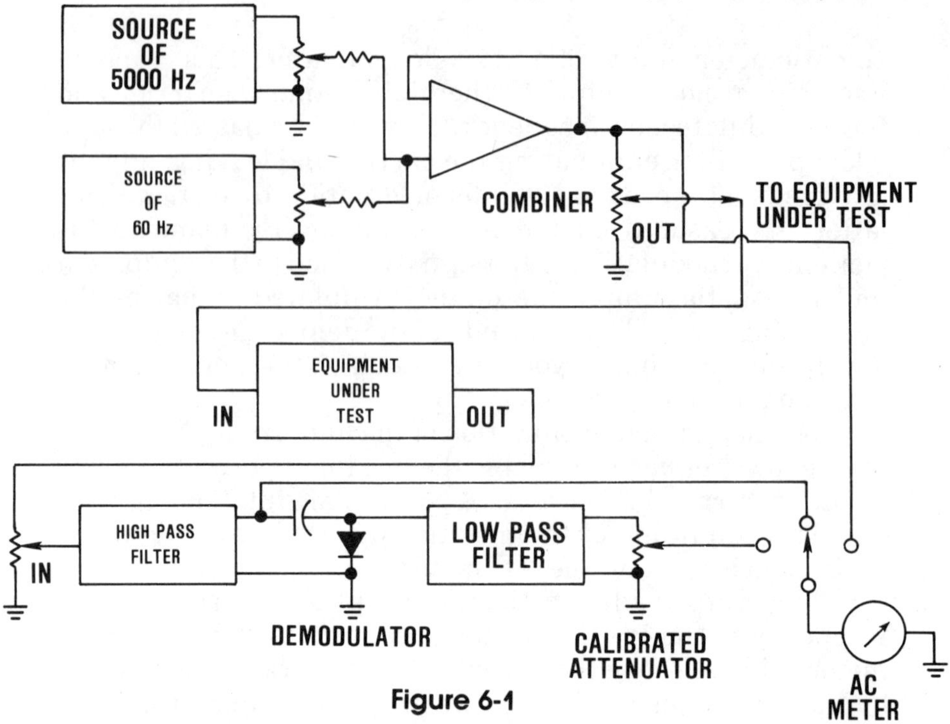

Figure 6-1

to provide an oscillation frequency of 5000Hz, the high frequency output signal of the device.

The low frequency 60Hz signal is derived from the secondary of the power supply transformer and is controlled by means of a 5K potentiometer. Noise on the 60Hz line is attenuated by a 2.2mfd shunt capacitor.

The two signals are added in the second half of the first 1458 op amp connected as a unity gain amplifier. Each signal is applied to the combining amplifier through a push-button switch and a decoupling 100K resistor. The push-button permits selection of either signal for amplitude and ratio adjustment.

The output of the combining amplifier is applied to a step attenuator. The voltage across this attenuator is switch selectable by means of the meter switch and may be read on the meter. This allows adjustment of the output level of the high and low frequency signals and therefore adjustment of their ratio. The attenuator provides 1, 0.1, 0.01 and 0.001 ratios to the output jack. This output is applied to the equipment under test through a standard phone jack.

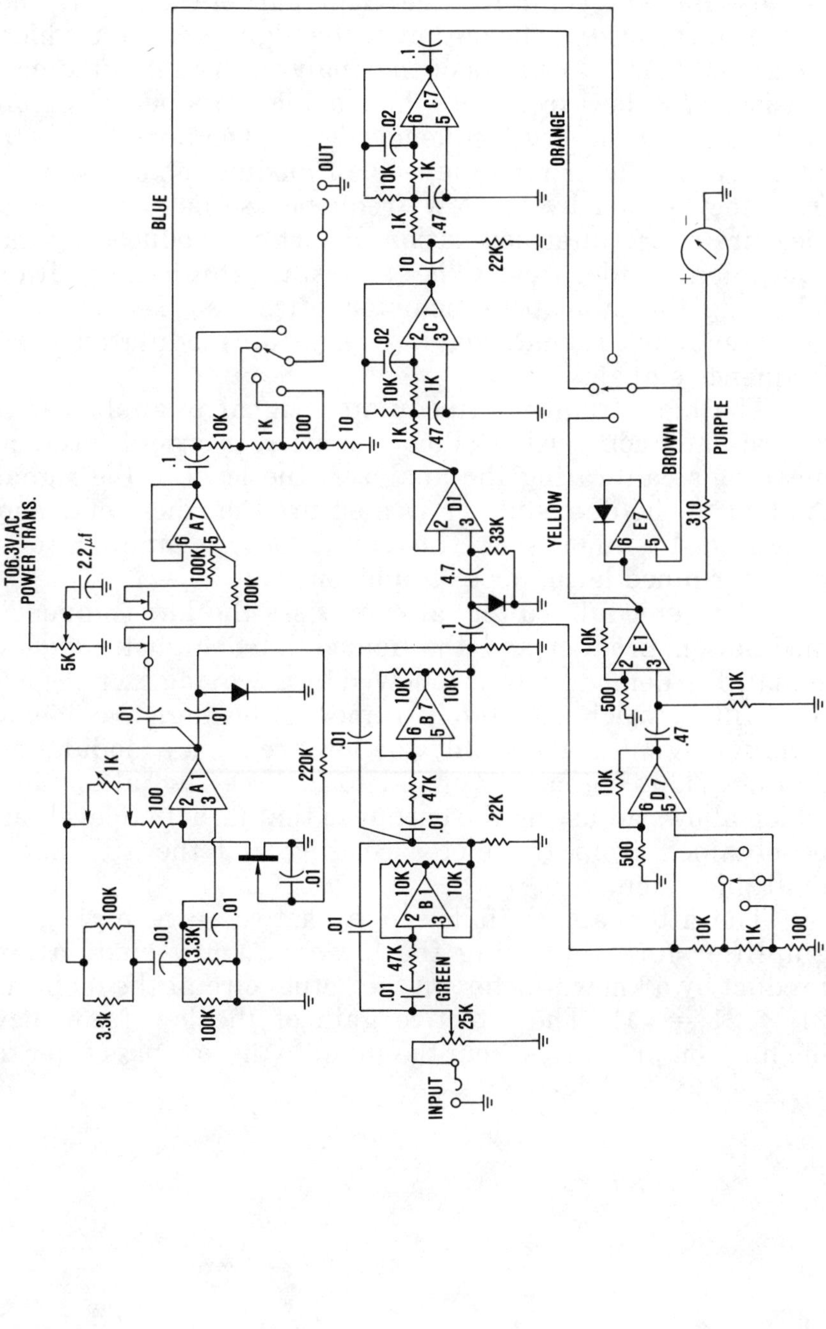
TO 6.3V AC POWER TRANS.
BLUE
OUT
ORANGE
YELLOW
BROWN
PURPLE
GREEN
INPUT

Figure 6-2

After the signal has been passed through the equipment it is applied to the input jack across which is connected a 25K potentiometer used for adjusting the signal level to a predetermined level in the analyzing part of the meter. The level is measured at the output of the high-pass filter which forms the first two stages of the analyzer. The input signal consists of a low frequency and a high frequency signal modulated by the low frequency signal. The purpose of the analyzer is to determine the level of modulation of the high frequency signal by the low frequency signal, which is a measure of the intermodulation distortion produced by the equipment under test. The two stage high-pass filter following the input potentiometer, effectively removes the low frequency signal, leaving only the modulated high frequency signal.

The high frequency modulated signal is applied to a demodulator comprising a diode rectifier. In order to provide means for calibrating the analyzer, the level of the signal applied to the demodulator is measured on the meter and adjusted by means of the input potentiometer to a predetermined level, for example, one volt.

The demodulated signal comprises the low frequency modulation product and the demodulated high frequency signal. The demodulator is followed by a two stage active low-pass filter which effectively removes the high frequency residual signal, leaving only the low frequency modulation product. This signal is applied to a three step attenuator which allows adjusting the output so that its actual level can be obtained from the meter reading and the attenuator multiplier factor.

The attenuator is followed by a two stage high gain amplifier for amplifying the low frequency modulation product by a known factor. In the actual circuit the factor is $21 \times 21 = 441$. The effective gain of the low frequency modulation product is from the input to the low-pass filter to the output of the high gain amplifier. The output of the high gain amplifier is measured on the meter.

The meter is an AC indicating meter using an active detector to rectify the applied AC. This circuit provides a linear response over a 10-1 range and the meter resistance is adjusted so that the meter reads 1 volt full scale.

Assembly and Wiring of an Intermodulation Distortion Meter

Figure 6-3 is a plan drawing of the circuit board of the intermodulation distortion meter. The board was a 4 × 6 ½ perf board with 0.1 inch hold spacings. Five 1458 dual op amps were used with five 8 pin half DIP sockets. The pattern of wiring indicates placement of all leads, although these leads are generally on the underside of the board. Resistor and capacitor leads that are long enough go directly to socket terminals, etc. Otherwise they are spliced with No. 24 bare wire. Leads that cross are insulated with plastic tubing.

All resistors are ¼ watt carbon. The .01 and .1 capacitors are ceramic disk. The .02 and .47 are metalized film printed circuit capacitors. The rectifiers are mini silicon rectifiers, except for the demodulator which is a mini germanium diode for better linearity at a signal voltage of 1 volt. This plan view also shows leads going to front panel controls and identifies them by color as well as function. All common ground connections are indicated by a ground symbol. Actually, a grounded wire was run around the under side of the board to facilitate these connections. To avoid confusion, this ground wire is not shown. Leads from components or other wires that go to socket pins are passed through a hole adjacent to the pin and soldered to the socket pin on the underside of the board as shown in Figure 6-4. This provides a good mechanical bond without undue strain on the delicate socket pin.

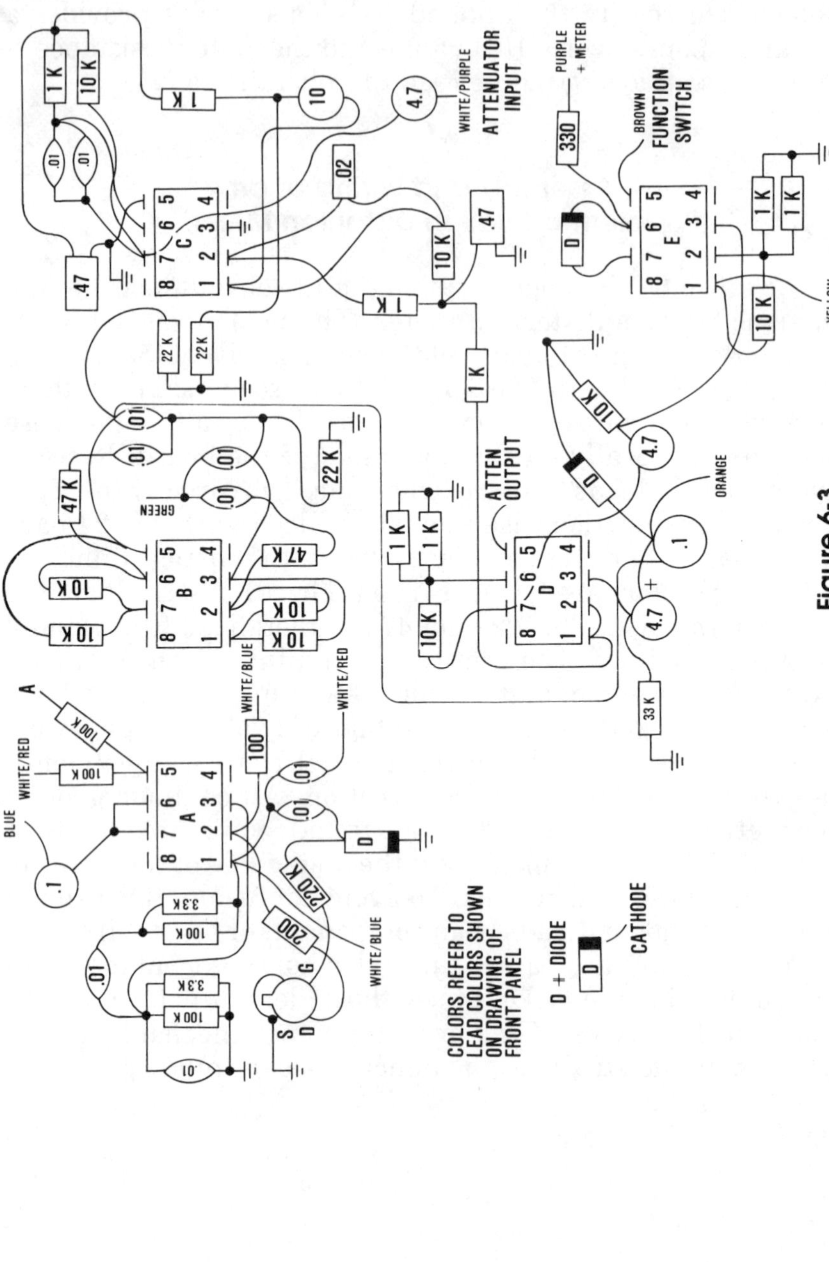
ATTENUATOR INPUT
WHITE/PURPLE
PURPLE
+ METER
BROWN
FUNCTION SWITCH
YELLOW
ATTEN OUTPUT
ORANGE
GREEN
WHITE/BLUE
WHITE/RED
BLUE
WHITE/RED
WHITE/BLUE
COLORS REFER TO LEAD COLORS SHOWN ON DRAWING OF FRONT PANEL
D + DIODE
CATHODE

Figure 6-3

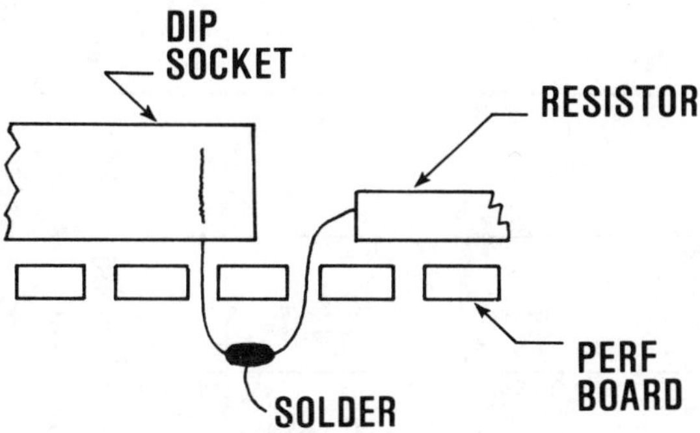

Figure 6-4

Mount the IM Distortion Meter in a Cabinet

A Radio Shack Cat. No. 270-0270 cabinet was of just about the best size and appearance for the IM Distortion Meter. The front panel was drilled as shown in Figure 6-5. The various controls, switches and meter were assembled and wired to the circuit board with color as shown in Figure 6-6. See also the photographs of the inside and of the front panel with all controls labeled.

Figure 6-7 is a photograph of the interior of the IM distortion meter and Figure 6-8 is a photograph showing the wiring at the rear of the front panel. Figure 6-9 is a photograph of the completed IM distortion meter. The power supply for the IM Distortion Meter is similar to the one used in the 3 watt amplifier (see Figures 2-1 and 2-3) except that the transformer is a 6.3 volt one, providing about ±8 volts of DC.

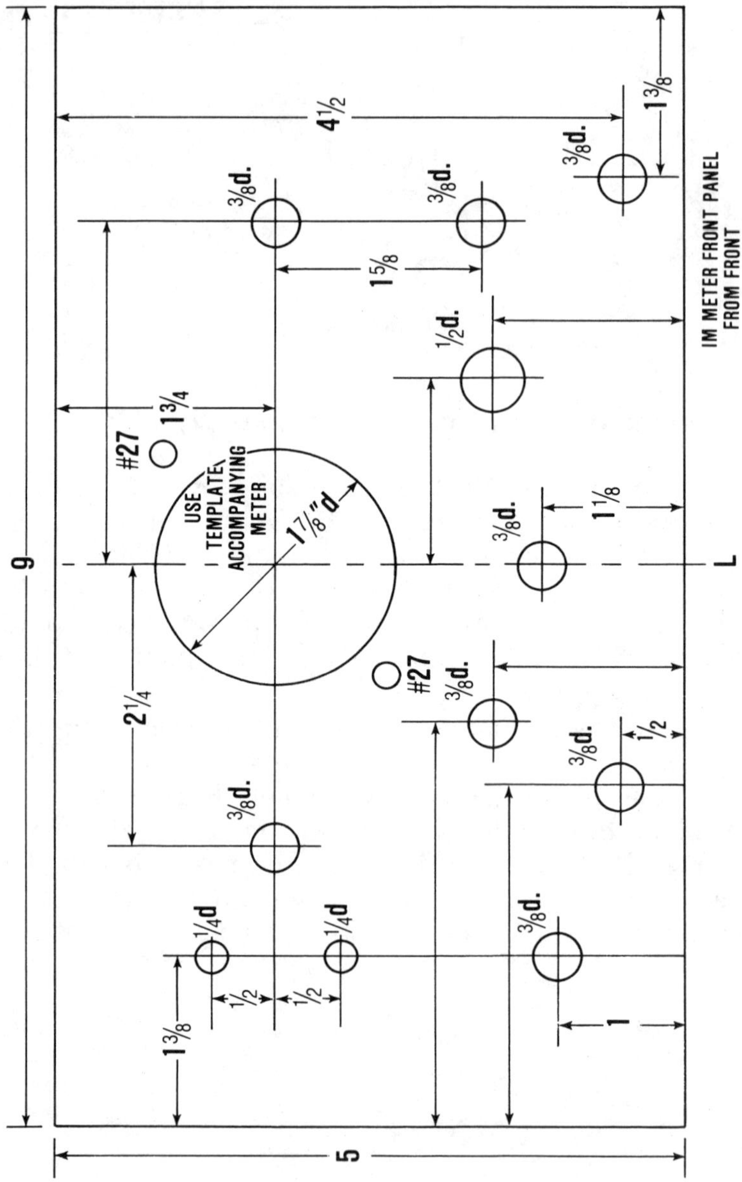

Figure 6-5: IM Meter Front Panel

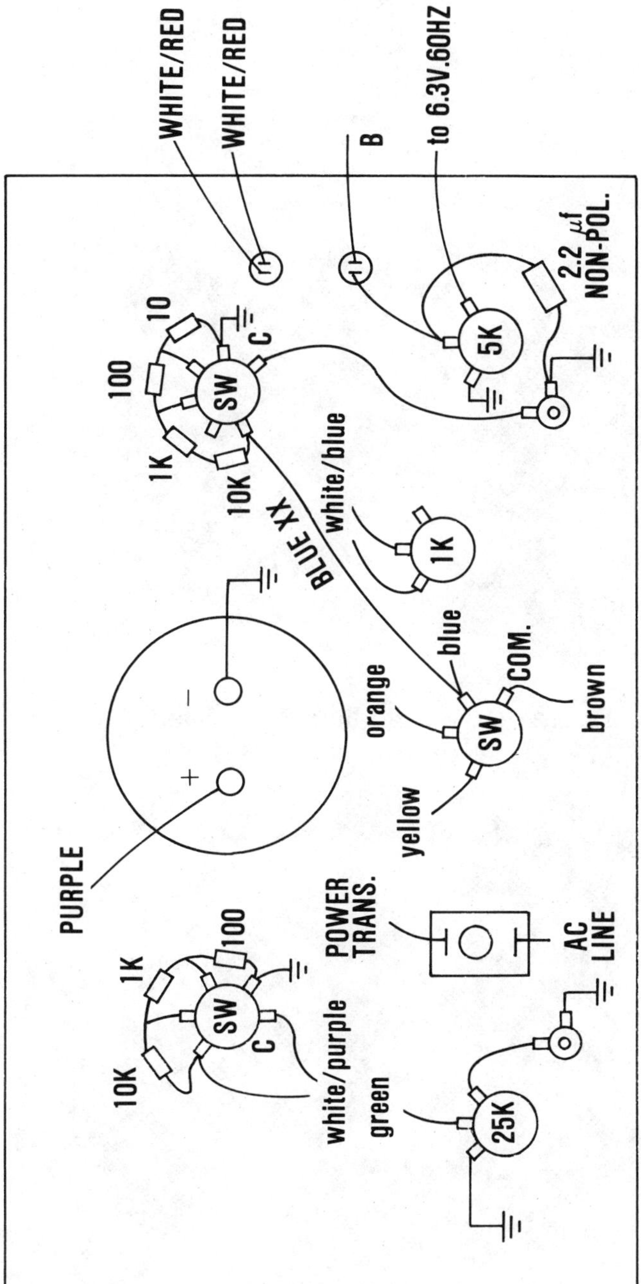

Figure 6-6: IM Meter Back of Front Panel Wiring

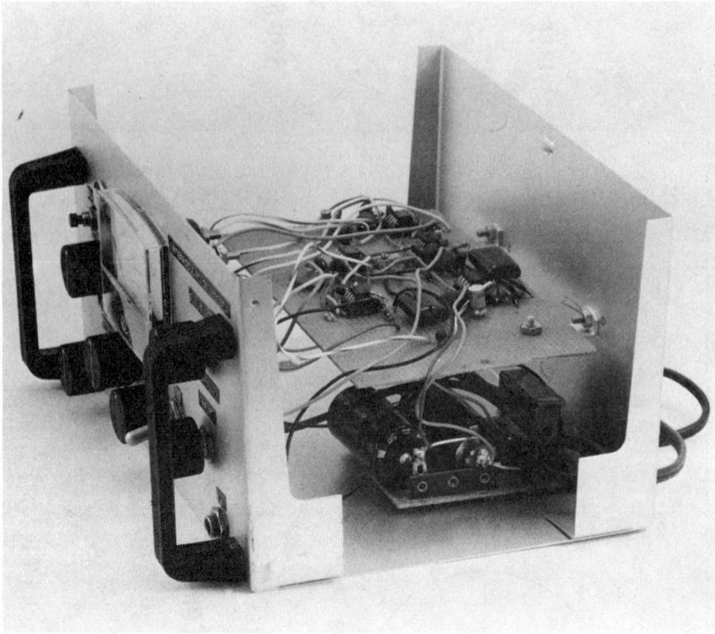

Figure 6-7: IM Distortion Meter, Interior

Figure 6-8: IM Distortion Meter, Rear of Panel Wiring

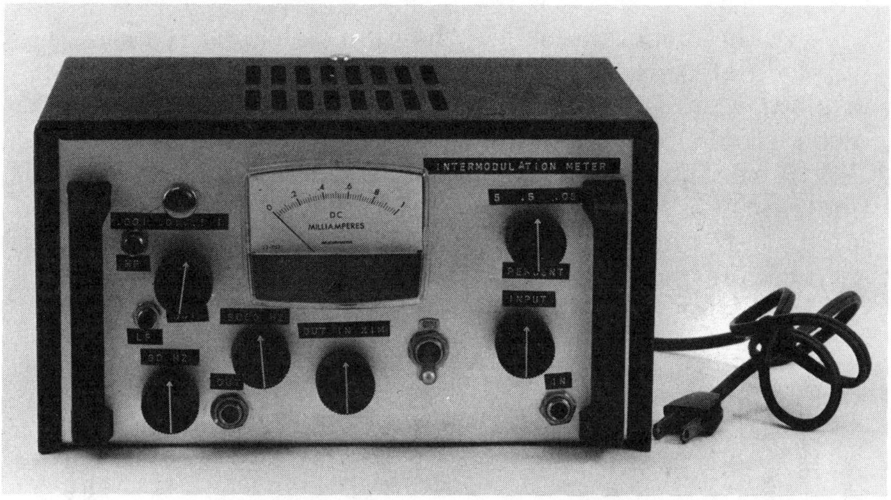

Figure 6-9: IM Distortion Meter

Calibrating the IM Meter

After the signal has passed through the high-pass filter, the signal comprises high frequency signal modulated by the low frequency signal. At this point the modulated signal is demodulated to separate out the low frequency signal. The following low-pass filter removes the residual high frequency signal, leaving only the low frequency signal as a measure of the intermodulation distortion. This low frequency distortion component is amplified and compared with the amplitude of the modulated signal before demodulation. It is this comparison which provides a measure of the percentage of intermodulation distortion.

Thus, if the amplitude of the modulated signal and the amplitude of the low frequency modulation component are known, the percentage of intermodulation distortion can be calculated. The modulated signal amplitude is measured on the output meter at the center position of the meter switch. The low frequency modulation component is measured with the switch in the right-hand position. However, the modulation component is multiplied by the low frequency gain between the low pass filter input and the output attenuator division factor (1, 10 and 100).

In the model described, the gain was 2000. Thus, if the unmodulated signal at the input to the low-pass filter is 1 volt, a .05 percent intermodulation distortion will read (.0005 × 2000 = .05) .1 on the meter with the output attenuator set at X1. A reading of 0.5 on the X10 attenuator setting would indicate .05 × 0.5 × 10 = 0.25 percent intermodulation distortion and so on. Figure 6-10 is a graph showing percentage modulation for various meter readings with a table of attenuator multipliers.

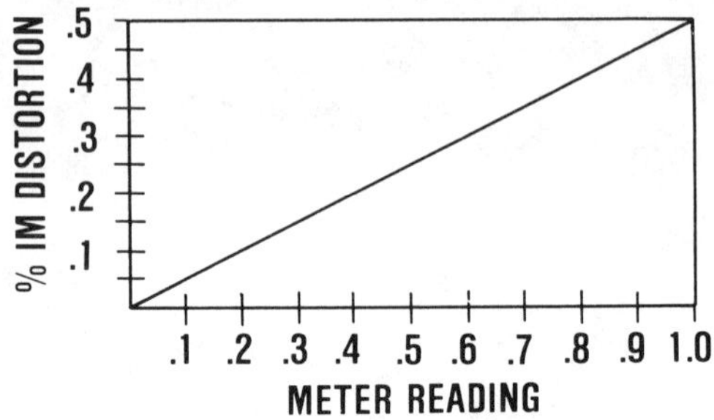

ATTENUATOR FACTORS

MARKING	MULT
.05	.1
.5	1.0
5.0	10.0

Figure 6-10

Measuring IM Distortion with the Meter

The IM Distortion Meter is connected to the equipment to be tested as shown in Figure 6-11. The output jack of the meter is connected to the input of the equipment to be tested and its output is connected to the input jack of the meter. The IM meter power is turned on. The indicating meter switch is set to "out." With the 60Hz potentiometer turned off (full counterclockwise), advance the 5000Hz control until an output is shown on the meter. Allow the circuit to warm up for a few minutes to stabilize its operation and then turn the 5000Hz control counterclockwise until the meter reads 1 volt (full scale). Press the HF button to cut out the 5000Hz signal and advance the 60Hz potentiometer until the desired percentage of 60Hz is shown on the meter. For the sake of this discussion, set it to 2 on the meter which will be 20 percent of the high frequency signal.

When the meter reads 1 volt, the actual output voltage is 1 volt multiplied by the output attenuator factors (1, .1, .01

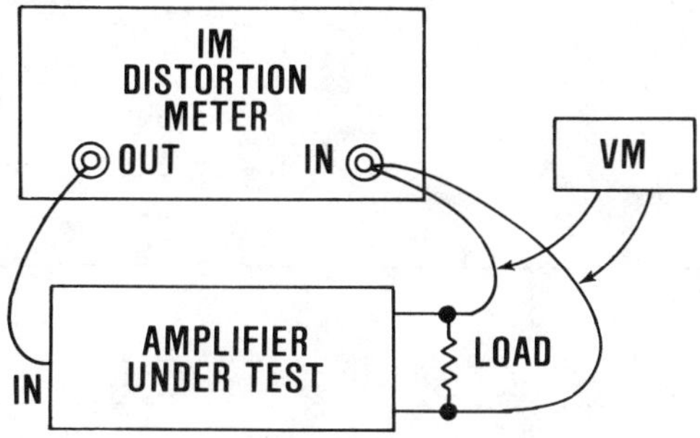

Figure 6-11

and .001). This output attenuator permits the application of appropriate voltages to the equipment under test. Further control of the voltage is given by the gain control of the equipment under test.

Figure 6-11 is a simple block diagram of a set-up for testing a power amplifier. The amplifier is loaded with a resistor representing its rated load, say, 8 ohms. The VM (AC voltmeter) across the load provides output voltage readings which are converted into equivalent watts output by dividing the square of the voltage by the value of the load resistance ($e^2/R = W$).

To make a run, the gain of the system is adjusted for some low output voltage, say, 1 volt. This would represent 0.125 watt with an 8 ohm load. The meter switch is turned to its center position marked "in" and the input gain control is varied until the meter reads full scale (1 volt). The meter switch is turned to its clockwise position marked %IM. The meter attenuator is set to a point where the meter reads its highest without going off scale, say, to .05. The percent IM distortion is then the meter reading in percent of full scale times .05, such as $.5 \times .05 = .025$, which is the percent of intermodulation distortion.

The process is repeated, increasing the amplifier gain and output reading of the IM distortion for each value of output voltage. With a good amplifier, the distortion will remain nearly constant up to a point at which it starts to increase sharply. This point of sharp increase is the overload power point of the amplifier.

Figure 6-12 is a plot of two IM distortion runs, one with the 3 watt amplifier and the other with the 30 watt amplifier.

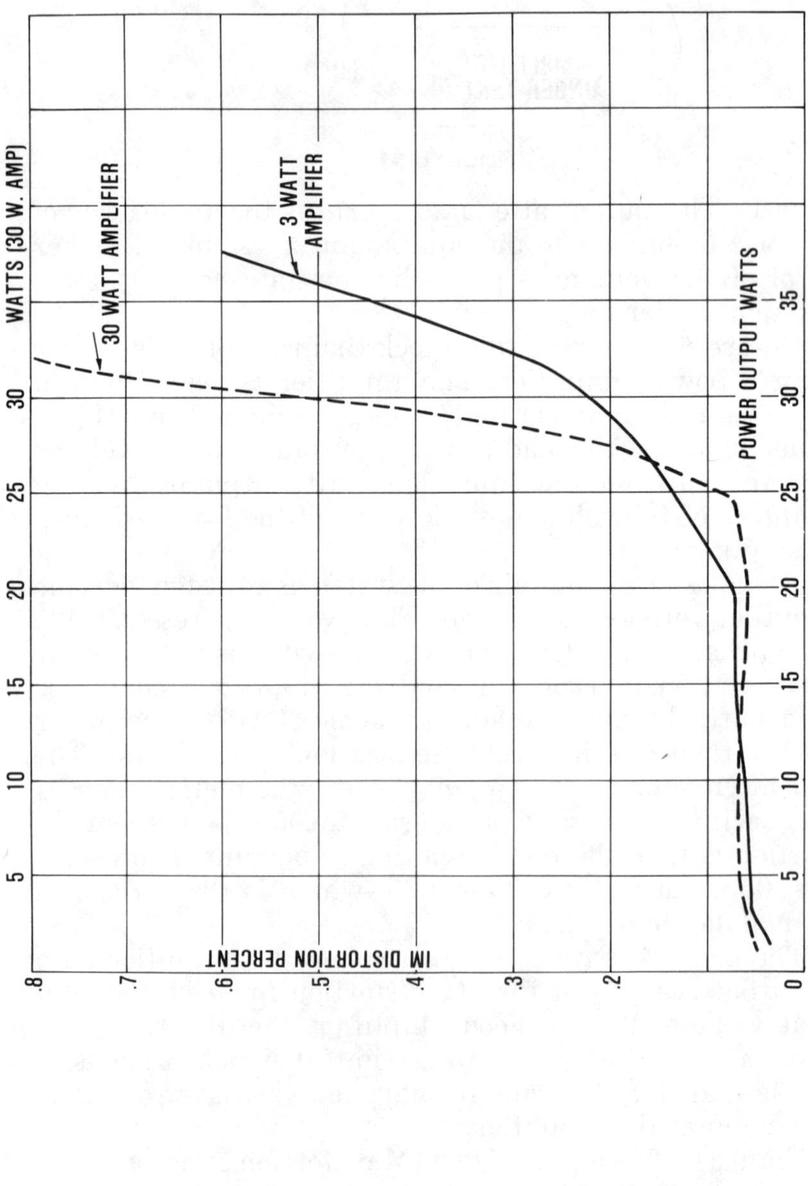
WATTS (30 W. AMP)
30 WATT AMPLIFIER
3 WATT AMPLIFIER
POWER OUTPUT WATTS
IM DISTORTION PERCENT
5 10 15 20 25 30 35
0 .5 1.0 1.5 2.0 2.5 3.0 3.5
.8 .7 .6 .5 .4 .3 .2 0

Figure 6-12

Taking 0.25 percent as the inflection point for each, the 3 watt amplifier starts to overload at 3.1 watts and the 30 watt amplifier starts at 27.5 watts.

These tests show three main characteristics of these amplifiers. First, there is no increase in distortion at very low levels that show that they have no significant cross-over distortion. Second, over a wide range, the IM distortion is very low, which shows a wide dynamic range of very low distortion. Third, these amplifiers show the definite points at which overload distortion starts.

Square Wave Testing Reveals Response Characteristic

One of the most revealing and yet easiest of tests to make on amplifiers, tape recorders and speakers is the square wave test. The ideal square wave rises and falls vertically and has an absolutely flat top. If the waves were analyzed for frequency components, they would extend from zero to infinity with definite predetermined phase relations between the components. When such a wave is passed through an electrical system, it is modified by the response of the system. The way in which it is modified tells a great deal about the system.

The technique used in square wave testing is to pass a square wave of known characteristics through the system being tested and to observe the output on a cathode ray oscilloscope. For a true picture of the system response, the characteristics of the square wave source must be known and the oscilloscope response must be substantially better than the response of the system being tested. Both of these requirements are not difficult to meet in most cases.

The flat top of the square wave is tilted more or less by the low frequency response of a system. The vertical rise and fall sides are tilted by the high frequency response, and ripple in the flat top is caused by peaks and valleys in the intermediate frequency range and phase shifting of these components. Overshoot on the leading edges is caused by high frequency emphasis and ringing.

Amplifiers having good characteristics naturally reproduce square waves rather well, as would be expected. Tape recorders and loudspeakers, on the other hand, are quickly judged by square wave testing which can show that they are deficient in a number of ways.

Build a Square Wave Generator

One of the most significant tests which can be made on a number of hi fi components is a square wave test. The only real drawback is that a cathode ray oscilloscope is required to interpret the tests.

Assuming that one has an oscilloscope or can borrow one for the tests, a simple and effective square wave generator will be described. A square wave response indicates not only the very significant transient response of a device or system but also has meaning in terms of frequency response. A square wave of a given frequency or repetition rate may be interpreted in terms of frequency response down to one-tenth the repetition rate and up to ten times the rate. Thus, a 500Hz square wave has significance down to 50Hz and up to 5000Hz, covering a large part of the audible frequency spectrum. (See page 215 for more on square wave testing.)

The circuit chosen is an overdriven RC oscillator with a period of 500Hz in turn over driving a buffer output amplifier. The circuit and constants are given in Figure 6-13.

These components were assembled and wired on a piece of perf board as shown in Figure 6-14. The IC op amp was a type 1458 dual op amp. A flasher on-indicator circuit was included, as recommended, as a reminder on all battery operated devices.

The circuit board assembly is shown in the photograph Figure 6-15 and the finished unit in the photo Figure 6-16. The unit was tested and found to provide a very clean square wave as shown in the photo Figure 6-17.

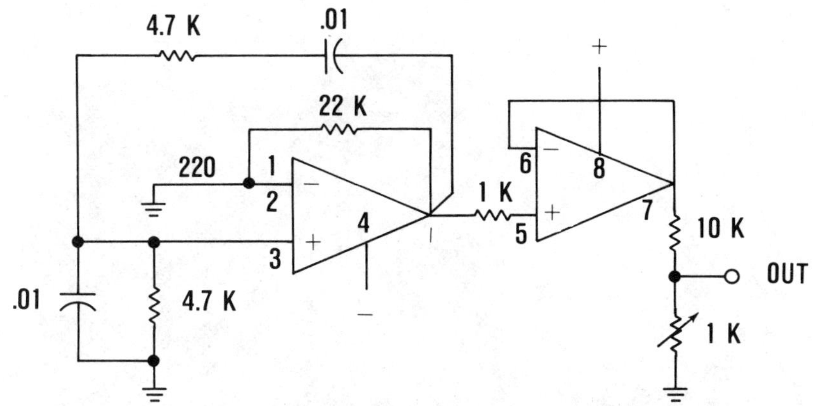

Figure 6-13

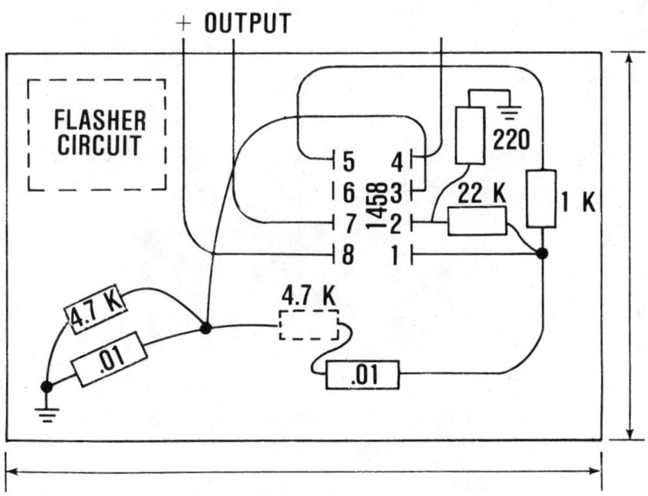

Figure 6-14

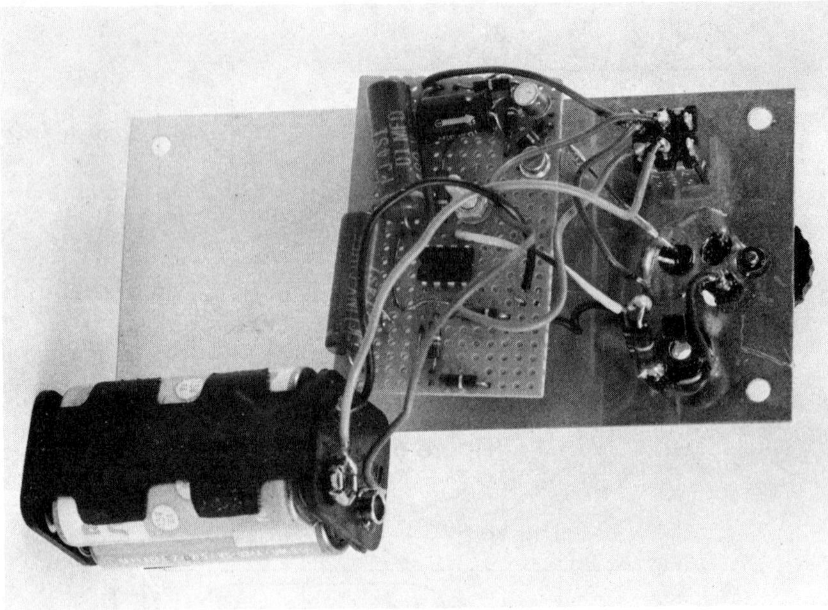

Figure 6-15: Square Wave Generator, Interior

Figure 6-16: Square Wave Generator

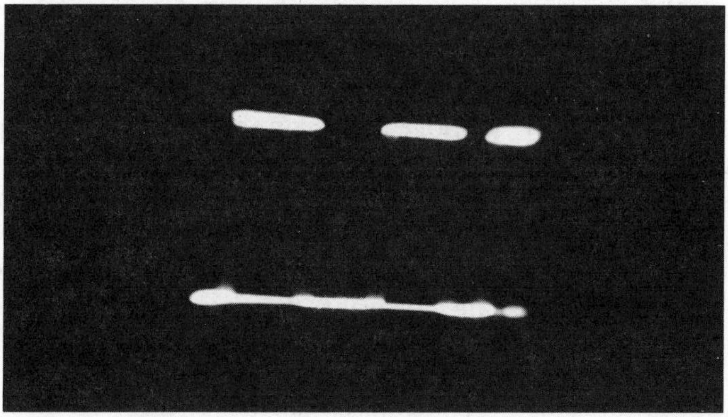

Figure 6-17: Square Wave on Cathode Ray Tube

Cathode Ray Oscilloscope—Don't Be Without One

Instantaneous visualization of audio voltages is one of the most revealing and interesting part of high fidelity experimentation. It can be done only with a cathode ray oscilloscope. The cathode ray oscilloscope consists in a television-like (picture) tube driven by at least two amplifiers. One of these amplifiers drives the horizontal plates (lines in television) and supplies a time reference, usually a linear sweep although it could be a reference sine wave. The second amplifier amplifies the audio signals and displays them vertically on the tube.

The display of audio signals on the cathode ray tube dance along with the music or speech in a never-ending parade. One quickly becomes fascinated with their scintillating waves. Sharp peaks denote transient bursts while slow undulations follow tremulous bass. The quality and texture of the sounds is traced in kaleidoscopic patterns. Sight adds a fascinating dimension to the sound.

On the technical side, the cathode ray oscilloscope is practically indispensable to the attainment of high fidelity sound. Steady state response of amplifiers, tape recorders and loudspeakers is easily carried out, although this can be somewhat time consuming. Harmonic distortion measurements require some rather sophisticated equipment, as do certain other aspects of audio fidelity, such as wow and flutter in a tape recorder. The results of such measurements are used as the basis for rating audio equipment. However, when it comes to the very important area of transient response, ratings in numbers become a real problem. This is where the cathode ray oscilloscope becomes vitally important. Not only does the cathode ray tube show up factors often overlooked, but it does this very easily and quickly. Such information is most important in the area of tape recorders and loudspeakers.

Tools for the Cathode Ray Oscilloscope

While it is visually engrossing to watch audio signals displayed on a cathode ray tube, technically, two basic signals should be observed. One, of course, is the sine wave covering the rull response range of the equipment involved, typically from 20Hz to 20,000Hz. Amplitude response (linearity) is easily traced. Overload levels may be observed. However, the oscilloscope that is receiving sine wave signals can only detect harmonic distortion when it reaches a rather high level in high fidelity terms, say, on the order of several percent. Good amplifiers will not show harmonic distortion in this way, although some types of equipment, such as tape recorders and loudspeakers, may.

The technical use of the oscilloscope is in the transient response department. A square wave is one of the most useful informative signals that has ever been invented. It is instantaneous, real time and permits continuous observation of a range of characteristics, even while changes and adjustments are being made. It not only describes transient response but also tells much about steady state. The 500Hz square wave generator described in this chapter will be used in the experiments to be described.

While distortion of a steady state sine wave may be difficult to observe, transient distortion is much more visible. It can be seen in tape recordings and becomes very conspicuous in loudspeaker performance. Loudspeaker performance is not easy to specify completely. Square wave response offers a welcome key to the need for an adequate means of rating loudspeakers.

Build a White Noise Generator

A "white noise" generator is defined as a source of random noise. That is, a noise which has equal energy distribution across the audio spectrum. Sometimes it is called shot or thermal noise. White noise is used to test loudspeakers and earphones. When white noise is applied to a loudspeaker or pair of earphones, it will have a smooth hissing sound, provided the frequency response of these devices is flat. Any coloration, ringing sound or the like indicates peaking of the frequency response. With this test, it is possible to adjust relative speaker levels, equalizers or tone controls to remove coloration from the response.

Like all listening tests, white noise testing requires experience and judgment which takes time to acquire. However, if one has the knack, white noise provides a very fast testing procedure.

There are a number of IC noise generators on the market. One of the easiest for one to obtain is the SN76477 as a "Sound Generator/Music Synthesizer IC" from Radio Shack (Cat. No. 276-1765). This device has many potential uses, such as sound effects for toys and synthesized music. However, it has a random noise generator as one of its basic elements. Unfortunately, the literature supplied with the unit does not clearly indicate how the noise source can be used as a white noise generator. With a little experimentation one can obtain a white noise output without other effects.

The connections to the IC pins and the external components used are shown in Figure 6-18. The noise filter connected to pins 5 and 6 is calculated to have a cut-off frequency well above audibility. The IC and external components were assembled on a piece of perf board as shown in Figure 6-18. The underside wiring is obvious in Figure 6-18. The numbered pin connections and circuit values are shown in Figure 6-19. The finished generator is shown in the photograph Figure 6-20.

Choose a Sound Level Meter Design

A sound level meter comprises a microphone, an amplifier and an indicator. If the indicator is a meter, some form of adjustable attenuator is required, if the meter is to cover a substantial range. Meters are practical for only a limited range, such as 10 to 1 (20dB). A useful sound level meter should cover 50dB or so.

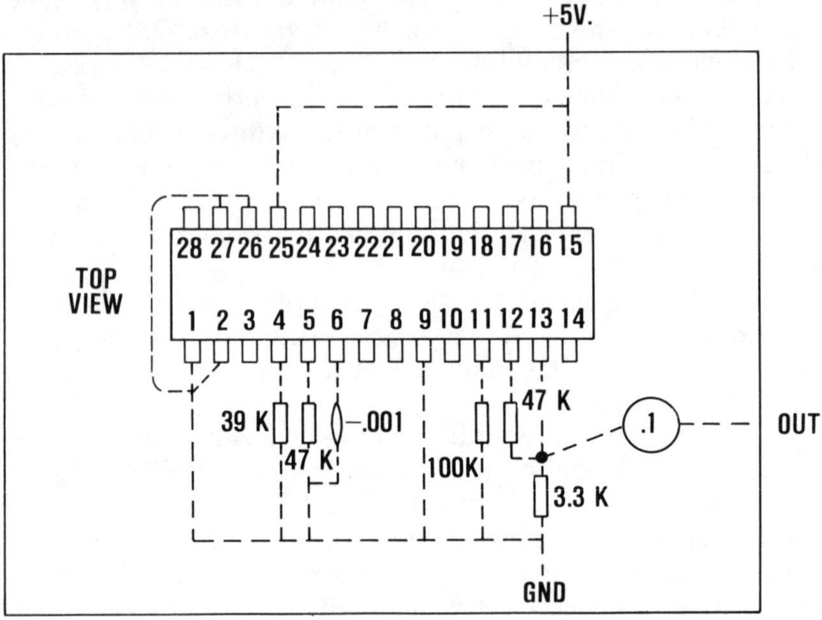

Figure 6-18

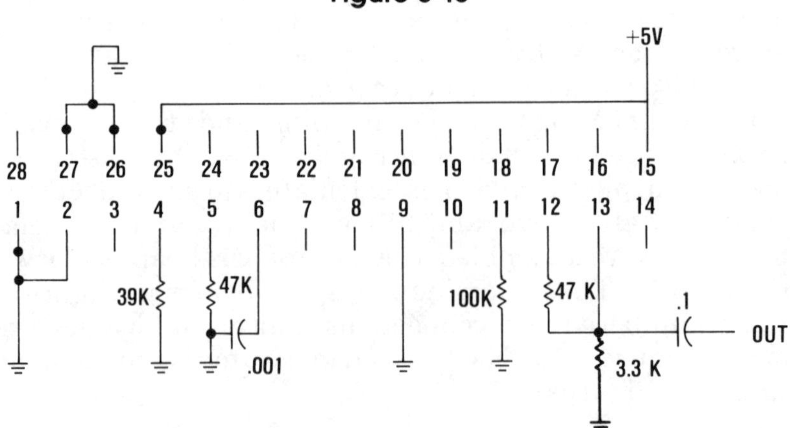

Figure 6-19

Figure 6-20: White Noise Generator

A succession of LED indicators can provide simple, inexpensive and highly practical sound level meter indicators. Each diode is programmed to light at a predetermined sound level in succession from the lowest level to the highest. No attenuator switching is required and the lights follow the sound level at high speed. Since this type of meter seems to be the more interesting, is easier and cheaper to build, it has been chosen for the Sound Level Meter Project to follow.

Breadboarding the Sound Level Meter

It is recommended that the sound level meter be breadboarded before final assembly. The circuit used is shown in Figure 6-21. You will get a feel for the circuit operation and be able to determine if the components are capable of performing the expected operation by breadboarding.

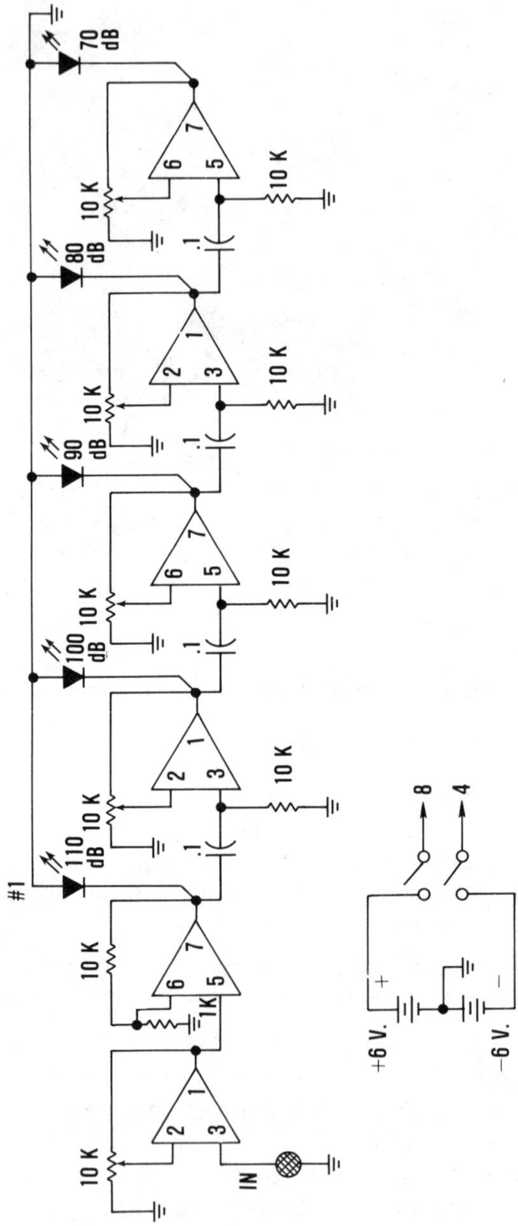

Figure 6-21

The first two amplifier stages are required to provide enough voltage to turn on the first LED indicator with an input sound level of 110dB. Microphone sensitivity is stated as the voltage in dBs below 1 volt which the microphone will produce at a sound level of 74dB. Thus, a microphone rated at -60dB will provide 1/1000 volt or 1 millivolt at 74dB.

The following table shows the rated millivolt output versus dB sensitivity ratings of microphones.

Microphone Sensitivity Table

Microphone Rating	Millivolt Output 74dB
-40 dB	3.2 mv
-50 dB	1.8 mv
-60 dB	1.0 mv
-65 dB	0.56 mv
-70 dB	0.31 mv
-75 dB	0.17 mv

The typical red LED which is recommended for this project starts to glow at around 1.5 DC, i.e., around 1 volt AC. Thus, it is possible to calculate the gain required to cause the first LED to glow at a sound level of 110dB. This 110dB level will provide 36dB or 63 times the voltage at the microphone than that shown above for 74dB. For example, a -70dB sensitivity rated microphone should supply 19.5 millivolts to the amplifier string input when impressed with a sound level of 110dB. This requires a gain of 1/0.0195 = 51 times in the first two stages to turn on the first (110dB) LED. If each succeeding stage is adjusted to have a gain of 3.2 times, the second through fifth LED will glow in succession at 100, 90, 80 and 70db respectively.

After experimenting with the breadboarded circuit to determine its practicality, it can be assembled and wired in permanent form.

Build a Fast Response Sound Level Meter

The sound level meter circuit was assembled and wired on a 1⅜ by 2⅜ inch piece of perf board with 0.1 inch hole spacings. The five gain control potentiometers were Radio Shack Cat. No. 271-218 printed circuit board type. They were mounted edgewise by drilling three No. 44 drill size holes for each pat. Connections were made on the underside. Three half DIP low profile sockets were used for the amplifiers. The board layout is shown in Figure 6-22. The wiring was completed as shown in the wiring diagrams of Figures 6-23A and 6-24B. The photograph of Figure 6-24 will help in completing this circuit panel.

The batteries were mounted in a Radio Shack Cat. No. 270-387 holder accommodating 8 "AA" batteries. One end of the battery string provided plus 6 volts, the other end minus 6 volts and a center-tap connection was connected to ground. A Radio Shack Cat. No. 270-233 phenolic box with an aluminum panel was used to house the meter. Five LED such as Radio Shack Cat. No. 276-041 or 276-033 were mounted through the panel by drilling snug-fitting holes and making solid with epoxy cement on the under side. The microphone was a Radio Shack Cat. No. 270-093 Dynamic Mike Element Omni Directional. It was mounted on the front panel with a piece of perf board to insulate the terminals and with epoxy to hold the assembly in place. Photograph Figure 6-25 shows the completed sound level meter.

After the wiring was completed, the unit was tested before assembly in the phenolic box.

Adjusting the Sound Level Meter

The sound level must be adjusted in order to correctly indicate predetermined sound levels. Use a calibrated sound level meter (rented or borrowed) for comparison. Set the two sound level meters equidistant from, but close to, a loudspeaker driven by a steady tone source, say, 1000Hz. Set the sound level at 110dB as measured by the reference meter. Adjust the 110dB potentiometer until the fifth LED just goes on. Reduce the sound level to 100dB and adjust the 100dB (second) potentiometer until the fourth LED just lights. Then continue at 90, 80 and 70dB, adjusting the respective potentiometers so that the corresponding LED just lights at the predetermined level.

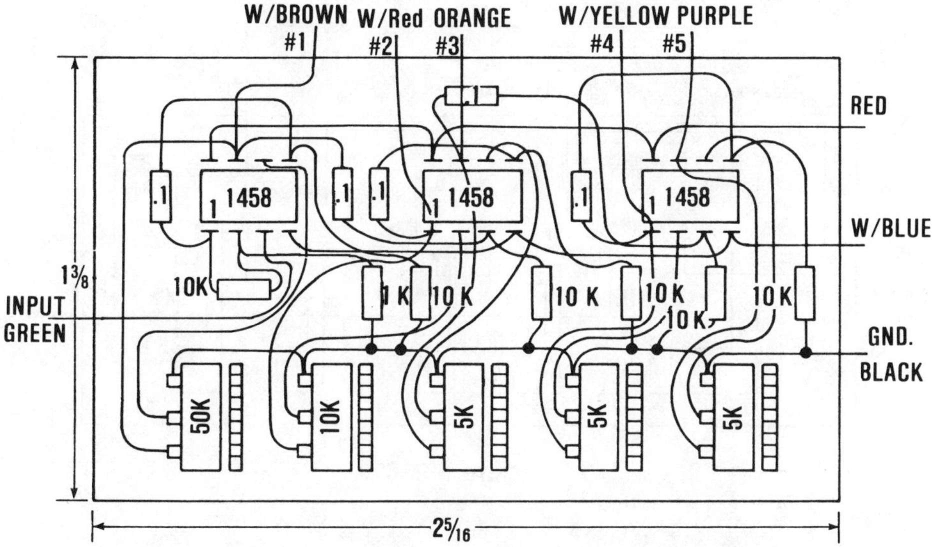

Figure 6-22

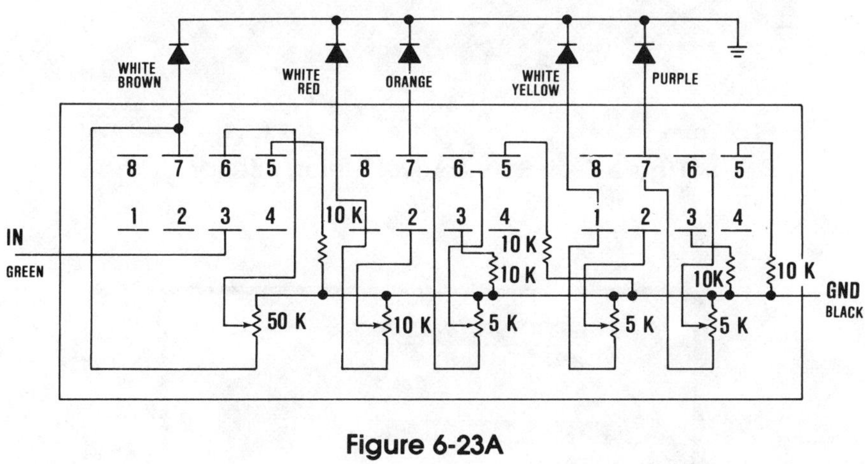

Figure 6-23A

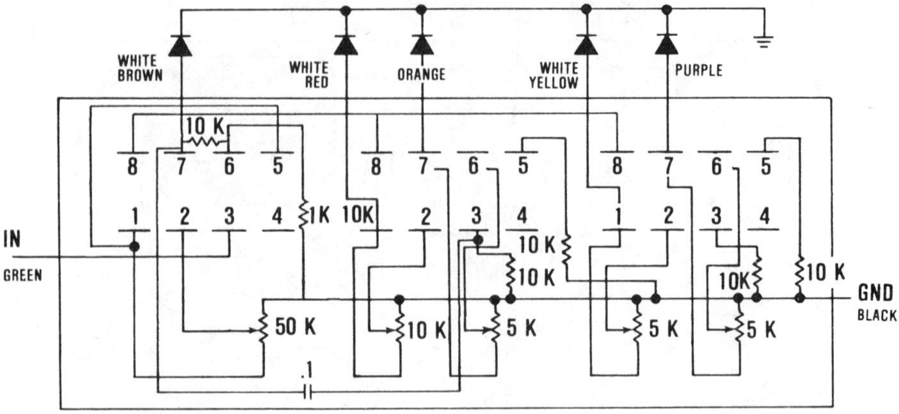

Figure 6-23B

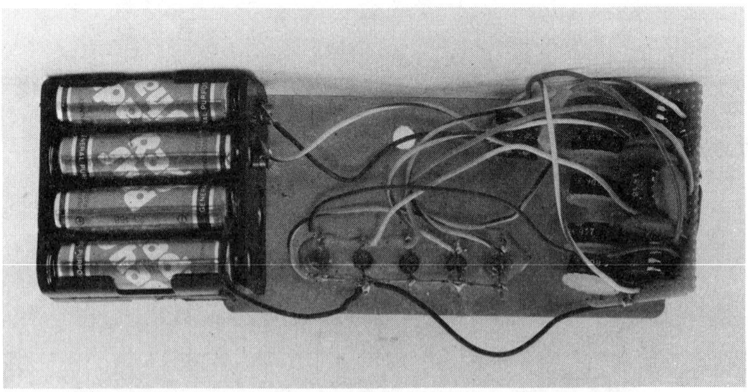

Figure 6-24: Sound Level Meter, Interior

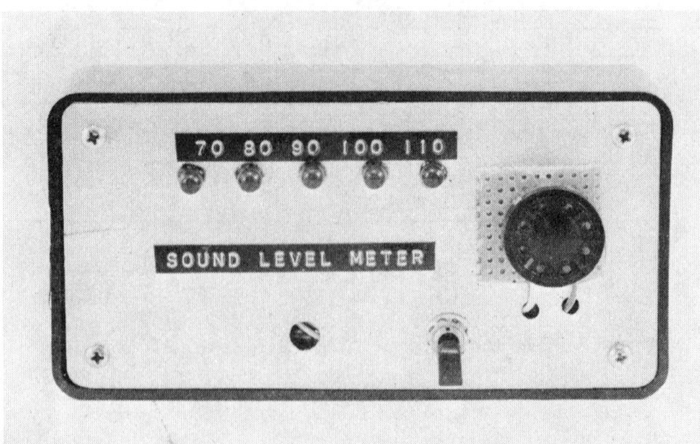

Figure 6-25: Sound Level Meter

Three Tone Test Oscillator

A test oscillator is one of the most useful tools for testing and evaluating components of a hi fi system. Perhaps the simplest yet informative type is a three frequency oscillator providing 100, 1000 and 10,000Hz. The 1000Hz tone provides a reference while the 100Hz is useful for evaluating low frequency response and 10,000Hz for high frequency response.

Choosing the Circuit

An op amp with a Wein bridge feedback is easy to build and can provide a very pure sine wave output. However, in order for the output to be a pure tone, the oscillator must oscillate at low level, particularly in a transistorized circuit. Typically, when there is feedback around a gain-providing device such as an op amp, oscillation takes place and builds up until the nonlinearity of the amplifier introduces losses which limit the oscillation amplitude. This nonlinearity introduces distortion in the waveform of the oscillator. One way to limit the oscillation amplitude while introducing a minimum of distortion is to provide the oscillator with an automatic level control which, in effect, reduces the feedback gain above a predetermined output level. These considerations led to a circuit including a Wein bridge feedback around an op amp, and with an automatic level control.

It was found that the tendency to oscillate changed significantly among the three frequencies and that the automatic level control was not enough to insure a pure sine wave output. Accordingly, a manual feedback gain control was also provided.

The circuit finally decided on is shown in Figure 6-26 which also shows the values of the components used. Although fixed values of capacitors and resistors with 5-10 percent tolerances, the plusses and minuses seemed to average out and the final frequencies were quite close to the chosen frequencies of 100, 1000 and 10,000Hz. A transistor driving an LED was added to show when the oscillator was actually oscillating. In operation, the manual gain control is advanced until the oscillator just starts to oscillate, as shown by the

LED indicator. Then the automatic level control takes over and holds the output steady.

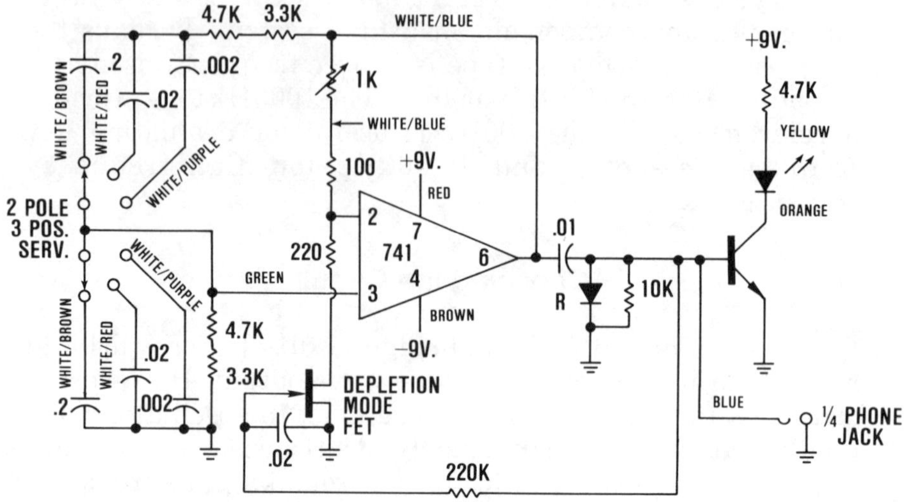

Figure 6-26

Building the Three Tone Test Oscillator

The circuit of Figure 6-26 was assembled on a $2\frac{3}{4} \times 2\frac{3}{4}$ piece of perf board as shown in the wiring diagram of Figure 6-27. The 0.2 mfd capacitors were small paper capacitors. The 0.02 and 0.002 capacitors were made up by paralleling two 0.01 and two 0.001 mfd small ceramic capacitors respectively. The automatic gain control FET was a Radio Shack Cat. No. 276-2035 depletion mode field effect transistor. The amplifier was a 741 op amp, and the LED driver a general purpose transistor such as a 2N2222. The rectifier was a small silicon diode. The wires going from the perf board to external parts such as the range switch were color coded for easy identification.

The power supply was a duplicate of the power supply used in the mixer panel and described in that connection. The complete oscillator is assembled in a cabinet (Radio Shack Cat. No. 270-253). The front panel layout is shown in Figure 6-28. The internal assembly, including the circuit board, is shown in Figure 6-29. The complete oscillator is shown in Figure 6-30.

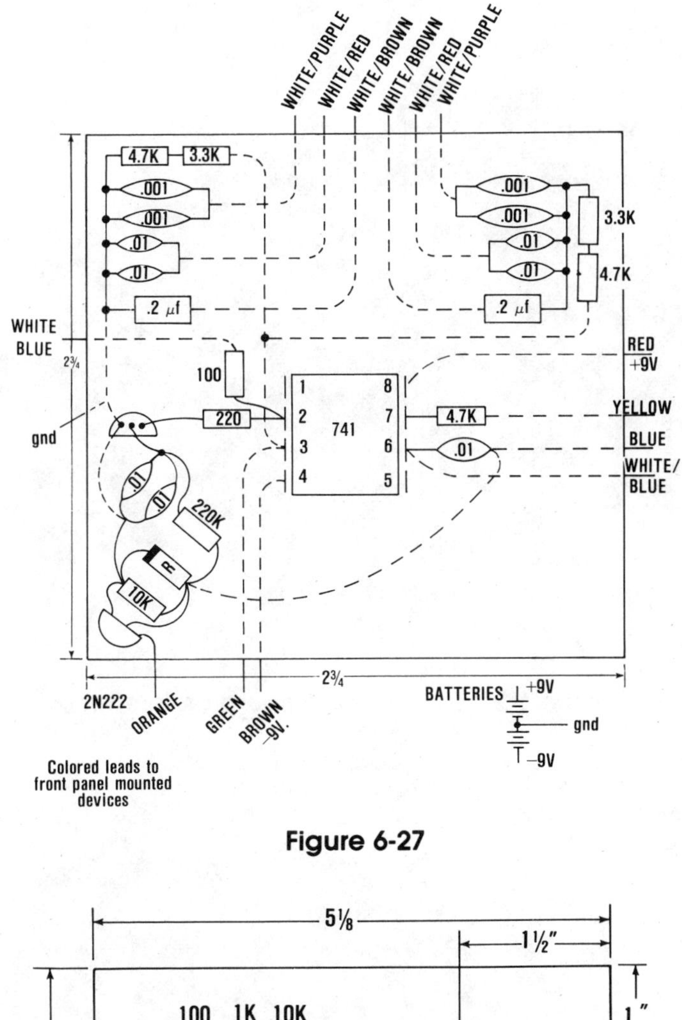

Figure 6-27

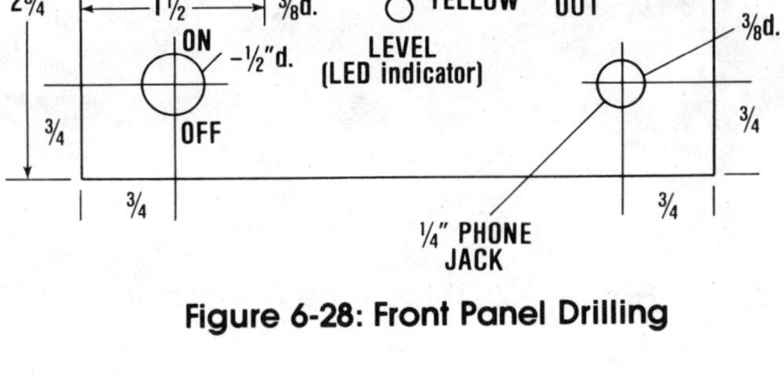

Figure 6-28: Front Panel Drilling

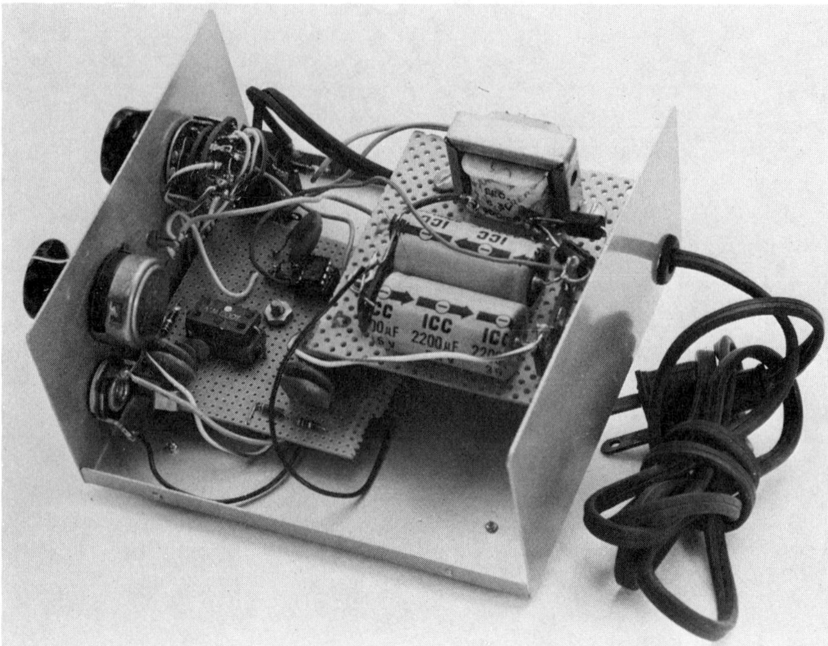

Figure 6-29: Three Tone Generator, Interior

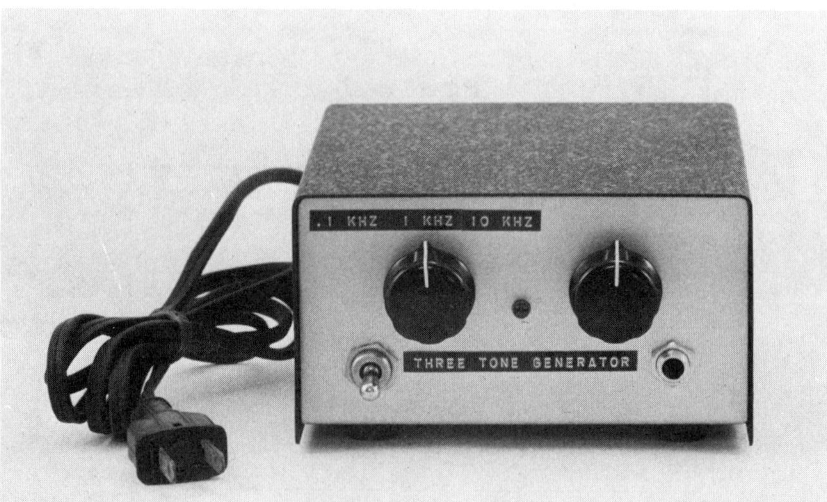

Figure 6-30: Three Tone Generator

Applying the Three Frequency Test Oscillator

The test oscillator was used in many tests for evaluating hi fi components as follows:

Loudspeakers may be tested using the Test Oscillator and the Sound Level Meter. Two way and three way speakers may be balanced. The balance and placement of stereo speakers is facilitated.

Audio amplifiers may be quickly checked for frequency coverage and for overload levels at the three frequencies.

Tape recorders present a particularly useful area for the test oscillator. Tape response can be quickly evaluated by comparing the 10kHz signal on playback with the 1kHz. Wow, flutter and rough response are easily detected using the 1kHz tone. Overload, particularly the high frequency saturation, may be determined with the 10kHz tone. The equalization required to boost the 10kHz response may be determined.

Build an Input/Output AC Meter

Sensitive AC voltmeters are useful in making measurements of performance in hi fi equipment. However, the usual VOM does not permit accurate measurements below a few volts. The experimenter can make a very useful AC voltmeter from a standard dB meter. These meters generally have full scale sensitivities of one volt or less and a frequency response covering the audio band. At the same time they have fairly high impedances.

The present project uses two dB meters to provide two AC voltmeters which may, for example, be employed as an input and an output meter in measuring the response of an amplifier or the like. Figure 6-31 shows how they may be connected. The 1K series resistors together with the cross-connected LEDs provide meter sensitivities of 1 volt at 0dB and overload protection. Making at least one of the LEDs across each meter visible through the meter case panel provides visual indication of overload.

To use these meters, first set the input at about 1kHz to read 0dB on the input meter and adjust the system gain to

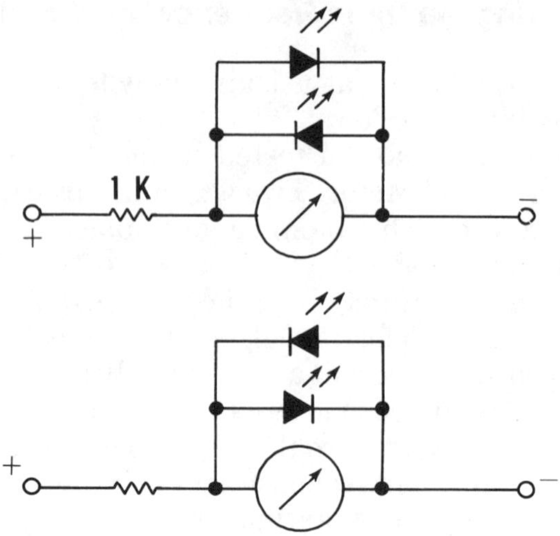

Figure 6-31

Figure 6-32: AC Meter

read 0dB on the output meter. Now change the input frequency, and if necessary readjust the input to again read 0dB. The reading of the output meter relative to 0dB will show the gain or loss at the new frequency in the system being tested. Repeat for other frequencies until the desired range of investigation has been covered.

The Radio Shack Cat. No. 22-dB meter is rated at 0.8 volt at 0dB with a frequency response flat from 20Hz to 20kHz and an impedance of 4000 ohms. The complete two-meter device is shown in the photograph Figure 6-32.

7

Testing and Trouble-Shooting Your Projects

Testing Your Projects

In the preceding chapters, operational projects such as a 3 watt amplifier, a 30 watt amplifier, a mixer panel, a 5 channel equalizer and a full size loudspeaker have been described in detail. Also a number of test instruments including a sound level meter, a square wave generator, a white noise generator, a three frequency test signal source and an intermodulation distortion meter have been described.

This chapter will show how to use these instruments and how to test the operational projects to obtain optimum performance and to detect and correct any malfunction.

Trouble-Shooting Your Projects

One real advantage of building your own projects is that you are in the best position to do any required trouble-shooting. Building the project should familiarize you with the circuits involved, step-by-step from input to output. If the project does not operate properly, the best way to find the trouble is to trace the signal step-by-step from input to output, including a check of both AC and DC levels. Point-to-point resistance checking is also necessary.

Step-by-Step Trouble-Shooting

Step 1. With power turned off, check all connections for poor soldering. Use a magnifying glass. Try moving wires to see if mechanically joined. Resolder if in doubt. Trace all circuits against the circuit diagram.

Step 2. Check the resistance to ground of both plus and minus bias lines. The normal resistance will depend on the circuit. When an ohmmeter is first applied to these circuits, it may read as a low value but gradually rises to several hundred ohms due to charging of filter capacitors. Any short, very low resistance or substantial inequality between the two resistances should be investigated. Look for shorts in the wiring. If an IC is suspected, disconnect the circuit showing the low value and check the IC terminal.

Step 3. Check to see that the plus and minus bias lines go directly to the appropriate circuit points. For example, in checking an amplifier using dual 1455 op amps, see that the plus line goes directly (zero resistance) to pin 8 and the minus line to pin 4.

Step 4. Check all resistive paths. For example, pin 3 in the circuit diagram is connected through a 10K resistor to ground. Check this with your ohmmeter.

Step 5. Turn on the power and check the plus and minus voltages to see that the power supply circuits are not being unduly loaded.

Step 6. Apply an appropriate signal to the input. Look for an output signal of the expected amplitude. When checking an amplifier, if the output does not appear normal, trace the signal stage-by-stage. The input signal should appear on the input pin of the first amplifier stage and in amplified form on its output pin. In checking an AC signal, a capacitor should be used in series with an AC voltmeter for this checking.

If operation does not appear normal, remove the input signal and check the DC levels on all input and output pins of the amplifier. An amplifier having substantial DC gain may have an abnormal input off-set, causing the output to saturate at plus or minus bias voltage. This must be corrected by replacing the amplifier or by decreasing its DC input impedance. At low signal levels, a small off-set on the output may create no problem but on high level stages it may limit the signal swing in the direction of the offset. When the output operates a DC responsive device such as a LED, it may turn the device on with no input signal.

When a trouble point is found it must be examined to determine the cause. Is it due to improper components, errors in connections, shorts or open circuits, a defective IC or even the wrong IC? Examine the circuit drawing and relate it to the mechanical circuitry in both form and function.

While an AC/DC voltmeter can be used, a cathode ray oscilloscope is a much more powerful tool in trouble-shooting. The scope shows things the voltmeters cannot show. For example, distortion, noise and overload provide information which a scope shows but not a voltmeter.

Test Your 3 Watt Amplifier

Frequency Response: The 3 frequency test generator provides three widely spaced test frequencies, namely, 100Hz, 1kHz and 10kHz. When connected to the amplifier as shown in Figure 7-1, this generator facilitates a quick check of overall frequency response. The generator shows a substantially constant output when the gain control is slowly advanced until the generator just starts to oscillate. The input control on the amplifier should be turned down until an output well below overload is obtained. The output at the three test frequencies should be almost the same since this amplifier is designed to be flat within a fraction of a dB over the test range.

Transient Response: The best and perhaps the only real way to test transient response is with a square wave. Tone bursts are an alternate method but they are more difficult to carry out. The square wave test can only be carried out with the aid of a cathode ray oscilloscope. If you have a scope to use or can borrow one, you can observe the transient response of your amplifier in the following manner. The simple set-up is shown in Figure 7-2.

First, connect the scope to the output of the square wave generator to observe the shape of its signal, a perfect "square wave." Then observe the square wave after passing through the 3 watt amplifier. It should be also a perfect square wave. The 3 watt amplifier should exhibit a substantially perfect transient response.

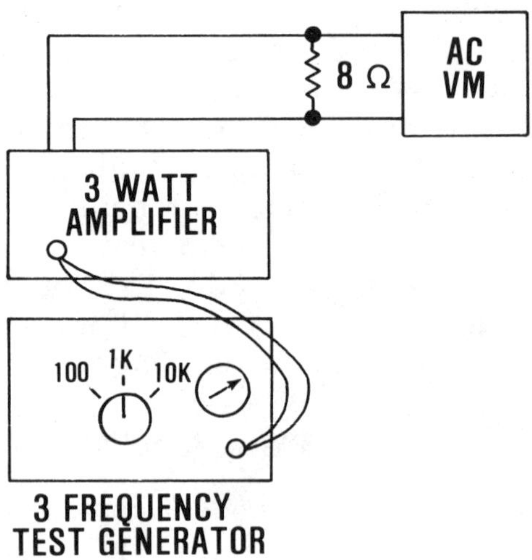

Figure 7-1

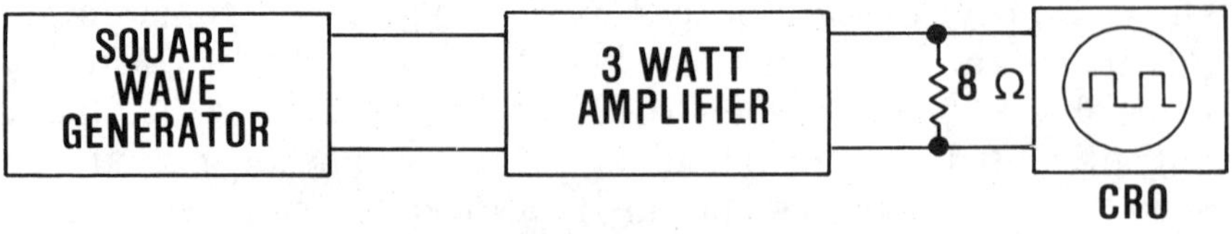

Figure 7-2

Intermodulation Distortion: Figure 7-3 shows how the IM Distortion Meter should be connected to measure IM distortion in the 3 watt amplifier. Measure IM distortion versus output as set forth in Chapter 6. The results should be in accordance with the graph of Figure 7-4. If the distortion is substantially different, repeat the trouble-shooting program outlined on pages 000.

Gain and Maximum Output: The 3 watt amplifier should provide maximum output with an input of 0.4 volt. Again, if it does not, repeat the trouble-shooting procedures. The maximum output is indicated by a sharp rise in IM distortion. This maximum output should be about 3 watts into 8 ohms. In a power amplifier, one source of distortion can be in serious mismatch between the output power transistors.

Measure Damping Factor

Damping factor is a term describing the damping effect an amplifier has on a reactive load such as a loudspeaker. Numerically, it is equal to the ratio of the load impedance to the internal resistance of the output amplifier. Factors of the order of 10-20 are supposed to be satisfactory. They may, however, in a high quality amplifier, be considerably higher.

To measure damping factor, set up an amplifier with a source of input signal and a switchable load as shown in Figure 7-5. If e_1 is the amplifier output voltage with no load and e_2 is the amplifier output voltage with the load connected, the damping factor is equal to $e_2/(e_1-e_2)$.

The 3 watt amplifier was measured and found to have a damping factor of 100—a very impressive figure.

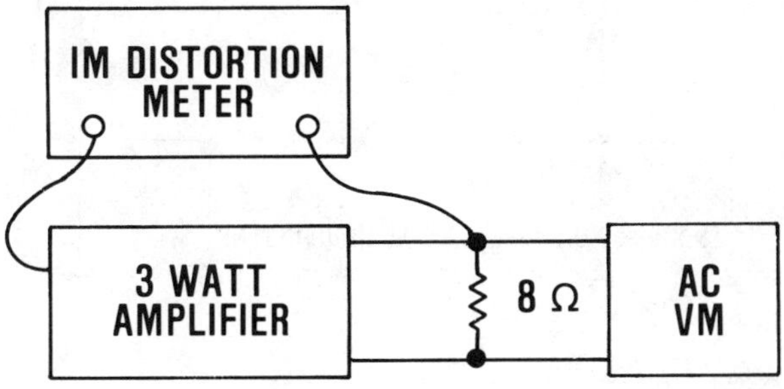

Figure 7-3

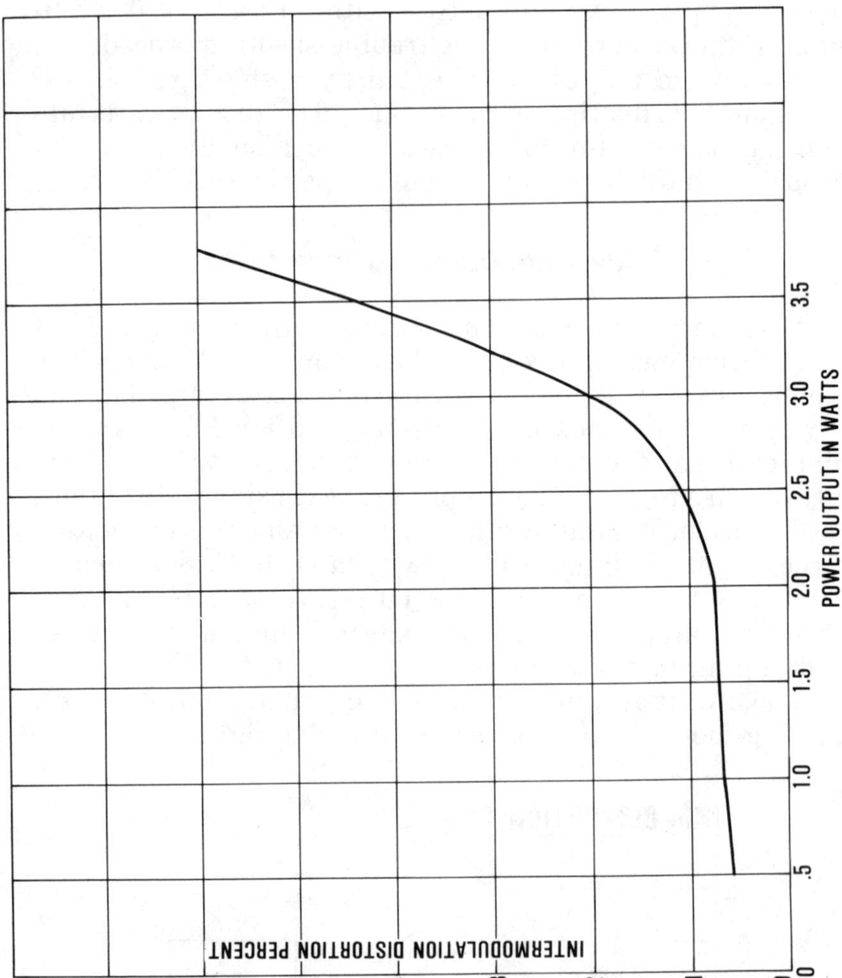

Figure 7-4: 3 Watt Amplifier IM Distortion

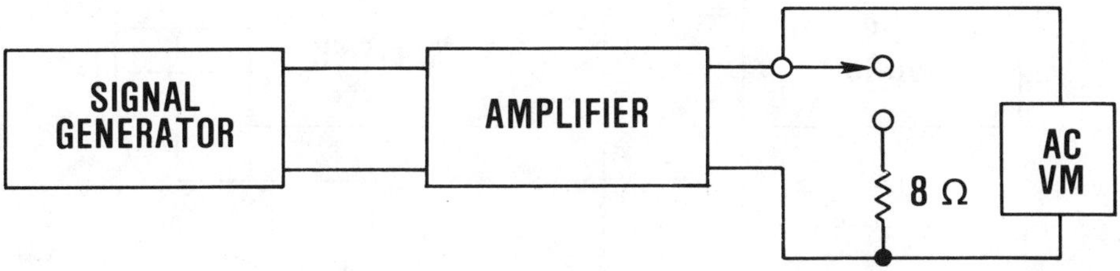

Figure 7-5

Measuring the Power to Your Loudspeaker

Alternating current electrical power into a load impedance is P = cos θ if θ is the phase angle between the current and voltage. This means that current times voltage is true power only when the phase angle is zero. The phase angle is zero for a pure resistance load. However, many devices powered by alternating current have inductive or capacitive components which cause the current to lag or lead the voltage. If this occurs, the power cannot be measured by a simple measurement of voltage and current, but only when the phase angle is known and is taken into account.

Thus, if one wishes to measure the true power into a loudspeaker one must know the phase angle of the speaker as a load impedance. Alternately, one can measure the voltage and current at frequencies at which the phase angle is zero.

First, one must find the frequencies at which the phase angle is zero and then measure the current and voltage at one or more of these frequencies. The set-up shown in Figure 7-6 was made using a source of variable frequency to drive the speaker. With the cathode ray oscilloscope horizontal and vertical gains adjusted to provide equal horizontal and vertical deflections of beam, zero phase angle is shown by a straight 45 degree line. Whenever the deflection opens to an ellipse or circle, a phase difference is indicated as illustrated in Figures 7-6a, b and c.

The "full size speaker" was tested in the same way. In-phase was found to occur at 140, 270, 800 and 10,000Hz and maximum out-of-phase at 60, 450 and 2000Hz. It was noted that the current tended to maximize at or near the zero phase frequencies. At 1kHz the phase was close to zero. Several

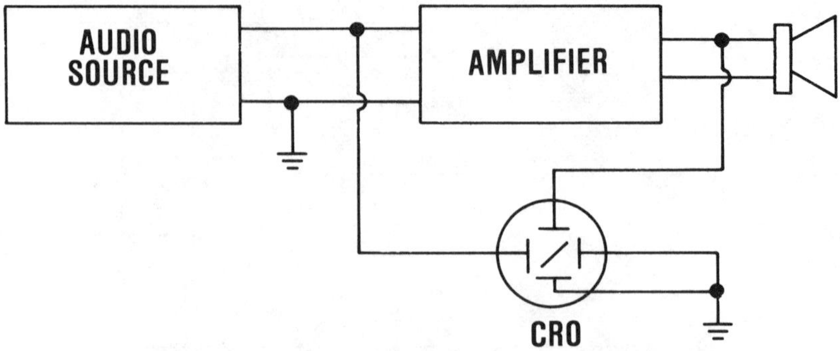

Figure 7-6

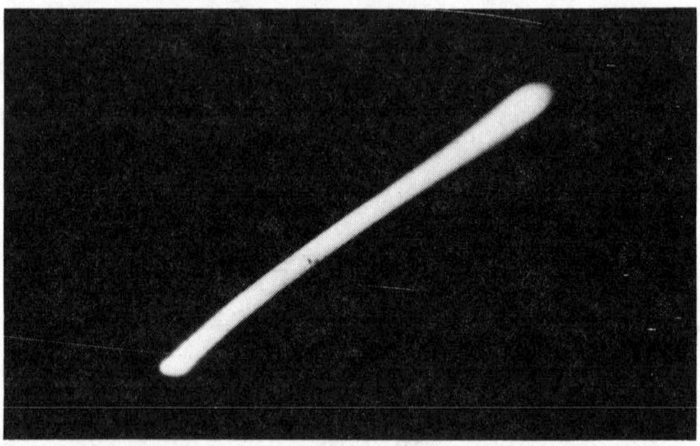

Figure 7-6A: Zero Phase Angle

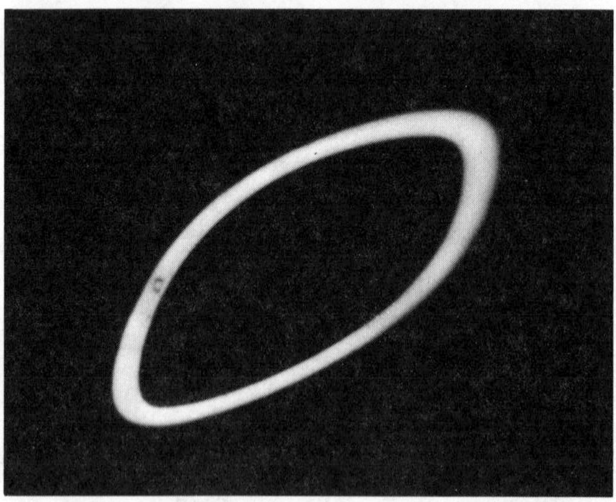

Figure 7-6B: 45 Degree Phase Angle

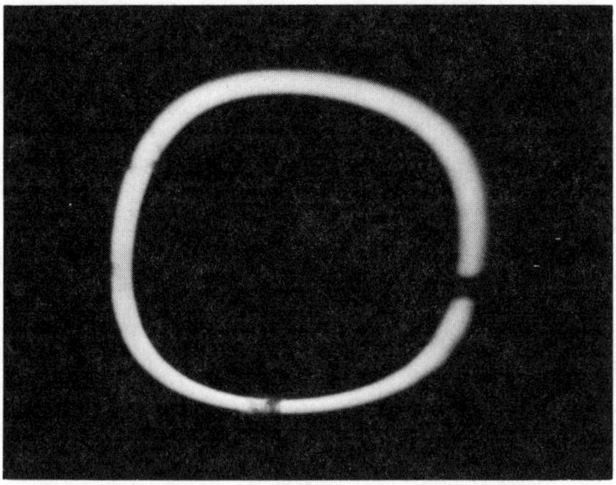

Figure 7-6C: 90 Degree Phase Angle

other speakers were tested. Some showed much more phase shift than others. One type, for example, drew nearly constant current from 20Hz to 20kHz and also showed little phase shift.

My conclusions were that while it is technically correct to make power measurements taking into account the phase angle between voltage and current, from a practical standpoint it is probably not worthwhile. Measurements at 1kHz are likely to be reasonably accurate unless there is some resonant effect that would be indicated by a dip in the measured current.

Based on the above tests and reasoning, it was decided that building a true power meter was not worth the trouble involved. Reasonably accurate measurements can be made without one.

Test Your 30 Watt Amplifier

Follow the same steps and procedures described above for testing your 3 watt amplifier to test your 30 watt amplifier. The IM distortion measured on the 30 watt amplifier is very similar to that plotted in Figure 7-4.

Test Your Loudspeaker

Your three tone generator provides three significant test frequencies for testing your loudspeaker. Using either your 3 watt or 30 watt amplifier, drive your loudspeaker from the three frequency generator. Use your Sound Level Meter placed about three feet from the center line of the speaker. While your Sound Level Meter has steps of 10dB, you can interpolate by observing the brightness of the LED indicators. If you find a serious unbalance, it is not difficult to reduce the response of one or more of the speakers by means of a constant resistance pad. Figure 7-7 shows the circuit of a two way speaker with an attenuator in the woofer circuit. If R is the impedance of the speaker, the pad is made up as shown with a resistance of 2 R in shunt and of R in series. The amplifier and cross-over network still see 8 ohms and the input to the speaker has been reduced 6dB. A similar network can be used as well in series with the midrange or tweeter speakers.

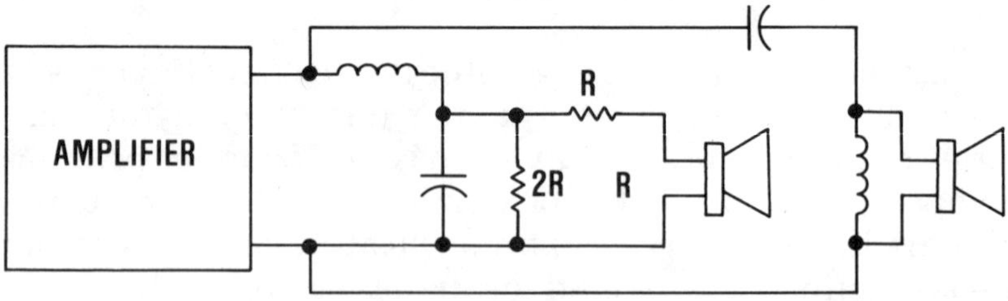

Figure 7-7

Figure 7-8 shows the constants for four pads providing 3, 6, 9 and 12dB attenuation respectively. If the speakers are to be driven by the 30 watt amplifier, the resistors should be at least 10 watts dissipation rating. If suitable 8 ohm resistors are not available, they can be made up of two 15 ohm resistors in parallel, or 10 ohm resistors may be used without materially affecting the results.

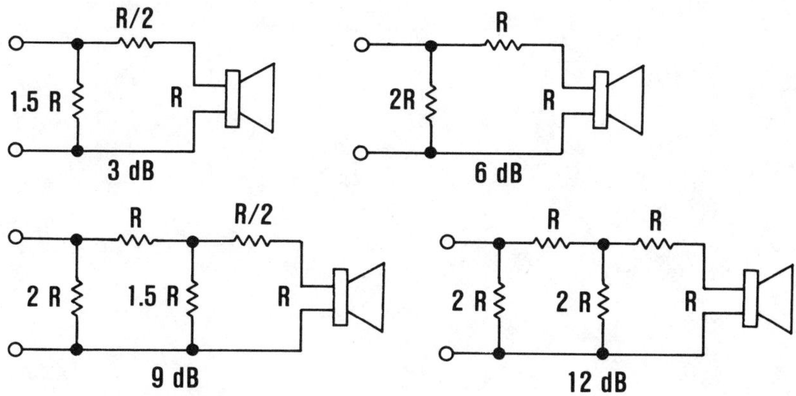

Figure 7-8

If one wants to be able to switch the pad in or out, the circuit of Figure 7-9 may be used, in which a double-pole double-throw switch cuts the pad in or out at will. Such a switching arrangement can be useful to provide some equalizing effect, for example, to de-emphasize the bass when the source is too boomy.

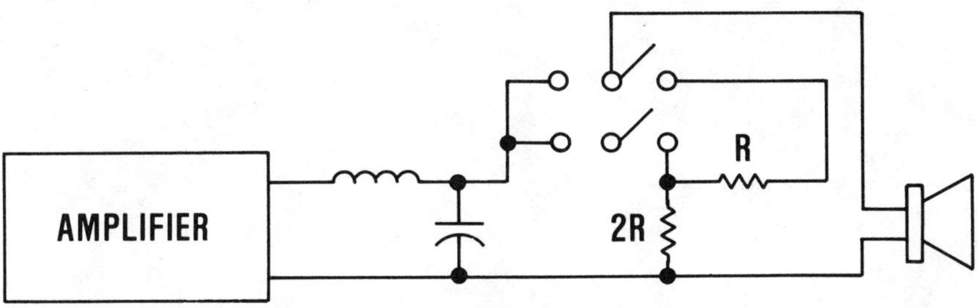

Figure 7-9

8

How to Select and Test Turntables, Tone Arms and Cartridges

Characteristics of a Good Record Player

The record player is another electromechanical component which merits careful consideration. It has three basic components: the turntable, the tone arm and the cartridge. We could add a fourth, namely, the needle. All have been the subject of a great deal of development and research and are available in many forms and at many price levels.

Perhaps some general observations can help set the stage. First of all, rather good turntables are available at moderate cost. Don't spend more than $200 unless you know what you really want. Second, if you are unfamiliar with products, start with the best known brands. Because of long experience and large production some of these make better products at substantially lower prices. Name brands are more likely to have spare parts and service readily available. Expensive gadgets may not be worth the trouble and expense and tend to malfunction.

Following is a brief discussion of the most important characteristics of turntables:

Dimensions. The dimensions do not vary materially, although if space is limited, they may warrant consideration. For example, a turntable on a high base may be inconvenient. On the other hand, dimensions have little to do with performance.

Weight: Turntable weights vary from as little as 7 pounds for the transport only to nearly 40 pounds for a turntable with tone arm and cartridge. Since a heavy table is important, make sure in purchasing a light machine that table weight is not sacrificed. A rugged base is also desirable, so the advantages seem to point to the heavier models.

Type: There are various types ranging from changers that play a stack of records, changing them automatically, to manual single players that play only one record at a time and must be completely serviced by hand. The changers are, of course, more expensive and have more of a tendency to get out of order. The single players require more attention. Variations such as automatic repeat are available. You must choose what type of performance best suits your needs. A semi-automatic which lifts the arm and stops the turntable at the end of a record is a viable alternative.

Speeds: The majority of turntables play both $33\frac{1}{3}$ and 45 rpm records. Not only the speed, but the record diameter, must be accommodated.

Speed adjustment: Many turntables include a speed adjustment of the order of ± 3 percent. This permits accurate adjustment of the speed which compensates for wear in the mechanism, or for adjusting a sharp or flat pitch to the music.

Motor type: There are many different types of motors used with turntables all the way from simple DC servos to coreless quartz-locked Hall. Synchronous motors are widely used. Specifications do not indicate superiority of one type over the other. Most types are capable of speed accuracies so nearly perfect as to be generally undetectable in any listening tests.

Drive type: There are two basic types of drives, namely, direct and belt. I don't feel these designations have much significance. Of course, if the belt breaks, one has the problem of finding a replacement. For performance and price, there is no clear indication of superiority or value one way or the other. At least some of the drive mechanism is common to both types.

Rumble: Rumble can be a very significant factor in a turntable. Rumble is the low frequency disturbance one hears due to mechanical vibrations in the system. Manufacturer's ratings range from -50 to -80dB (weighted DIN B standard). The type of installation may modify these ratings. For example, a soft pad under the turntable may reduce rumble while a hard shelf installation may increase it. A figure around -70dB would seem to be a good place to start looking.

Wow/Flutter: Wow and flutter are the high frequency disturbances due to rapid fluctuations in turntable speed. You can reasonably assume that factors of less than 0.1 percent are undetectable to the ear.

The following items are related to the tone arm and cartridge characteristics. Some features are essential while some are gilding the lily. Your budget may help make decisions as to which features are practical necessities. Some features add convenience while others are designed to improve reproduction at a price. Some features are arm parameters while others apply solely to the cartridge. The tone arm and cartridge, however, should be evaluated together. For example, it would be an error to combine an incompatible cartridge with a highly sophisticated tone arm.

What to Look for in a Tone Arm

Vertical Tracking Force (VTF): A high-compliance cartridge needs a lower VTF than a low-compliance cartridge. The cartridge needle must follow the record grooves accurately. If the needle has a low compliance, it pushes harder on the cartridge and tone arm and if the resistance of the tone arm is too low, it reacts by reducing the net motion of the electrical signal generator in the cartridge and, hence, the resultant response. On the other hand, too high a VTF makes for greater wear on the record. On the more sophisticated tone arms, VTF is adjustable from 0 to about 4 grams. The VTF should be adjusted to match the cartridge requirements.

Antiskating: The grooves on a record spiral inwardly creating an inward pressure, which, if unopposed, causes wear on the inner wall of the record grooves. This is countered by the antiskating mechanism. Again, this is adjustable on sophisticated arms from 0 to 4 or so grams.

Mass: As the needle follows the grooves across the record it must move the tone arm. Zero mass in the tone arm would minimize record wear but would not provide resistance to needle movement in following groove undulations, thus preventing output from the record. A high mass would provide maximum output but also maximum wear on the record. Low mass is one compromise, but this makes the system more vulnerable to external mechanical disturbances such as when music is played for dancing.

Tracking Error: One of the anomalous situations in disc recording is that the recording is done with a stylus which moves so that it is always tangent to the groove, and the playback stylus often does not. The playback cartridge in most cases is mounted on an arm which is pivoted somewhere beyond the edge of the turntable. It thus describes an arc across the record that is truly tangent at only one point—at about the center of the arc. The result is pressure on the sides of the grooves that cause wear and tear and damage to high frequency components of the recorded sound, and also increases noise. Curving the tone arm and/or angling the cartridge decreases the error angle to true tangency. Making the tone arm longer also helps. Reducing the error to the order of one degree is considered acceptable. A few expensive players are built with means for moving the playback cartridge across the record to maintain true tangency. The practical question remains, is perfection here worth the price?

Features: Many features are available in tone arms. Some are quite important while others are a matter of personal preference. Dust covers are recommended since dust is a source of trouble. Shock absorbent feet are useful but can be substituted for by other soft mounting. Other features that can be evaluated on a personal basis are: LED strobe, optical speed indicator, logic control, provision for two tone arms, cuing (selection of predetermined parts of a record), and so on, all at widely varying prices.

Aids in Selecting your Cartridge

The most important component of any record player is the cartridge. The cartridge includes the needle which supposedly follows the record grooves faithfully and generates an output voltage accurately representing the grooves. Ideally, the needle moves with zero mass and infinite compliance and communicates this movement to its signal generating component without loss or distortion. This also includes critical damping, which means no overshoot or ringing. Zero wear on the record should be a requirement of the ideal cartridge. Since the cartridge is such a vital component and since the ideal has never been attained, let us examine the practical answers. There are a multitude of cartridges with varying degrees of perfection available at a wide range of prices. In order to place things in proper perspective, it should be mentioned that few records are available on the market which warrant the highest degree of development in cartridges. However, if you are very selective and are willing to wait for improvements, you may be justified in insisting on the best cartridge available. On the other hand, the best cartridge costs a lot more than what may be practically adequate.

Following are comments on the principal characteristics of phonograph cartridges:

Type: There are three principal types of cartridges, namely, moving iron, moving coil and moving magnet. Other types and variations include, electret, crystal, ceramic and ribbon. While the output of these cartridges is typically 2 to 6 millivolts, some, including the electret, are significantly lower. Low output invites problems such as noise and hum pickup.

Stylus: There is probably no item for which there is more variation. The stylus should be wide enough so that it does not ride on the bottom of the groove generating noise, yet narrow enough to accurately follow the groove side walls. The most common form is elliptical, on the order of 0.3 × 0.7 mils. Refinements and special forms include such designations as "Stereohedron," "Quadrahedron," "Shibata," "Long-contact hyperbolic," "Biradial elliptical," and so on. It is not evident what these designations connote or what advantages they offer, but it seems that there is general agreement that an elliptical form is preferable.

Vertical Tracking Force (VTF): The vertical tracking force is generally specified and usually lies between 0.75 and 3 grams. This should be within the range provided by the tone arm. Higher VTF may indicate older designs which unnecessarily wear records.

Output: Output voltage is an important characteristic of cartridges since the balance of the system must be compatible. If the cartridge output is less than the normally expected level, additional amplification may be necessary and inconvenient. A normal range is generally considered to be from 2 to 5 millivolts. Higher outputs, of course, are no problem; much lower ones may be solved as pointed out previously.

Response: Frequency response is one of the most important characteristics of a cartridge, yet one which is not easily defined. Since cartridges are electromechanical devices, they exhibit complex responses. Simply to say a given cartridge has a response from 20Hz to 20kHz may leave much undefined. A cartridge may have a serious peak below 20Hz which emphasizes rumble and low frequency noise in records. It may have a high frequency response above which response falls rapidly. It may have a transient response which is difficult to describe. Damping may be too much or too little for best reproduction. Harmonic distortion may be excessive. Before choosing a cartridge, obtain as much technical information regarding its performance as possible and make studied comparisons including price, unless price is no object. In the latter case, of course, choose the one which when completely specified promises the best overall performance.

Separation: Separation describes the distinction maintained between the two channels of a stereo signal. It generally varies between a low of 20dB and a high of 35dB. With other things being equal, the higher the better, but there is no magic number here.

Load: Generally phono amplifiers are designed to provide a 47K load with a capacitance ranging from around 275 to 450 pf. Cartridges requiring substantially different loading can be padded to make up the difference.

Features: Any features which seems nonstandard or would require special accommodation should be carefully considered as to whether or not they are desirable for your particular use. Any unusual claims should be substantiated.

Integration: The turntable, tone arm and cartridge must cooperate in an integrated system. Each component must be compatible with the others so that they all operate harmoniously together. For example, a low mass tone arm should be provided with a low mass, high-compliance cartridge. Otherwise an advantage gained in one element may be swamped by some incompatible characteristic of another.

Finding Good Records

Your record player can never be better than the records you play on it. Record making is a complex art with many contributing factors. The process starts with the source of the sounds to be reproduced, the microphones that pick up this sound, the number and placement of these microphones, the mixing of the microphone outputs, the production of the master tape or disc, the production of the master die, the material of the final record and its pressing. These factors are modified by the skill and judgment of those involved, the quality of the various machines involved, and an overriding cost factor. You must be the final judge of the end product.

Your personal tastes dictate the type of music you desire. They may also dictate the performers, record company, etc.

This makes it difficult to make selections on the basis of someone else's recommendation. Records cannot be rated solely on technical grounds.

Some record companies, however, seem to turn out better records on the average than others. These companies are the ones to look to for the better records.

Judging a record should be done in much the same way one should judge loudspeakers (see page 000). Avoid warped records. Store records in a manner that avoids warping. Protect records from dust and heat.

The most promising sources of high quality records are often the smaller companies, especially those using direct-to-disk and digital mastering techniques. They get a higher price for their records which is warranted by the care used in the processing.

Record care is important. Keep records very clean. Even old records can often be brought back to life by proper cleaning.

Using Test Records to Fine Tune Your System

The only scientific way to test your record playing system is with a test record. A number of these are available, but they are not cheap and are not always easy to obtain. These records generally provide a variety of tests covering most of the essential characteristics of a record playing system.

Following is a list of the tests provided by The Stereo Review Model SRT-14-A stereo test record:

1. Cartridge tracking, high frequency.
2. Frequency response, 20kHz to 25Hz.
3. Separation, left-to-right.
4. Separation, right-to-left.
5. Cartridge tracking, low frequency.
6. Broad-band noise for channel balance.
7. Cartridge and speaker phasing.
8. Low frequency noise.
9. Turntable flutter.

10. Frequency-response sweep, 500-20,000Hz left channel (requires chart recorder, meter or oscilloscope).
11. Frequency-response sweep, 500-20,000Hz right channel.
12. Intermodulation distortion (requires IM distortion analyzer).
13. Antiskating adjustment.
14. 1000Hz reference tones (requires oscilloscope).
15. 3,150Hz tone for flutter and speed accuracy (requires flutter and/or frequency meter).
16. Stereo spread.
17. Standard "A" 440 Hz.
18. Chromatic octave.
19. Guitar-tuning tones.

Test records come with detailed instructions on how to carry out the various tests and how to evaluate your record player in various areas of performance.

9

How to Get the Most Out of Your Tape Deck

Notes on Magnetic Tape Recording

Magnetic recording goes back to the Poulsen telegraphone, a wire recorder invented in 1900 by Valdemar Poulsen. This machine recorded on steel wire, but its use was severely limited because amplifiers did not exist then, nor for some time after its invention.

Around 1930, German researchers developed a recording method using paper tape coated with iron oxide. Brought to this country after World War II, it was rapidly developed along several lines. Plastic disks and plastic tape quickly replaced the German paper tape. Good amplifiers were available and the magnetic recorder became useful and widely applied.

Early applications of magnetic recording were mainly limited to speech. However, the idea of recording music as well, began to develop. In this area, there was a serious problem. The magnetization curve of iron, steel and iron oxide has the familiar S-shape and is only linear over a limited range. First attempts were made to DC bias the recording head to the center of the linear range. This was only partially successful.

For all practical purposes, at least until the present time, Marvin Camras of the Armour Research Foundation solved the linearity problem. He demonstrated a recording process in which a high frequency bias current is added to the recording current. The bias current causes the recording current effect on the magnetic surface to skip the nonlinear toe portion of the magnetization curve. (See Figure 9-1.) The result is that the recording is produced in the linear regions of the magnetization curve in both positive and negative directions. On playback the bias current disappears due to the limited frequency response of the system leaving the desired signals.

While this was a big step toward low distortion magnetic recording, the results were far from perfect. The so-called linear part of the magnetization curve is not exactly straight. Hence, the greater the excursion of the recording signal, the greater the harmonic distortion will be. At the other extreme is noise that is generated by the finite size and random distribution of the particles that make up the magnetic coating. Reducing the recording signal to reduce distortion decreases the ratio of signal to noise (S/N). The range is thus limited on the one hand by noise, and on the other hand by distortion. This is another way of saying that from a practical standpoint, magnetic tape has a limited dynamic range. If 3 percent distortion is used as the upper limit, a dynamic range of the average tape is on the order of 60dB from noise to overload.

High Frequency Response of Magnetic Recording

Since the playback voltage induced in the pickup head is proportional to the rate of change of flux induced in it by the recorded pattern, the playback voltage is proportional to frequency. (See Figure 9-2.) However, since the pickup head has a finite gap, there will be some high frequency at which the response flattens out and then falls. There are three ways to help this situation. One is to use a narrower head gap, and another is to move the tape faster so that at a given frequency a wave length covers a greater distance. The third is to add frequency compensation in recording or playing back or both. This latter is called equalization (EQ).

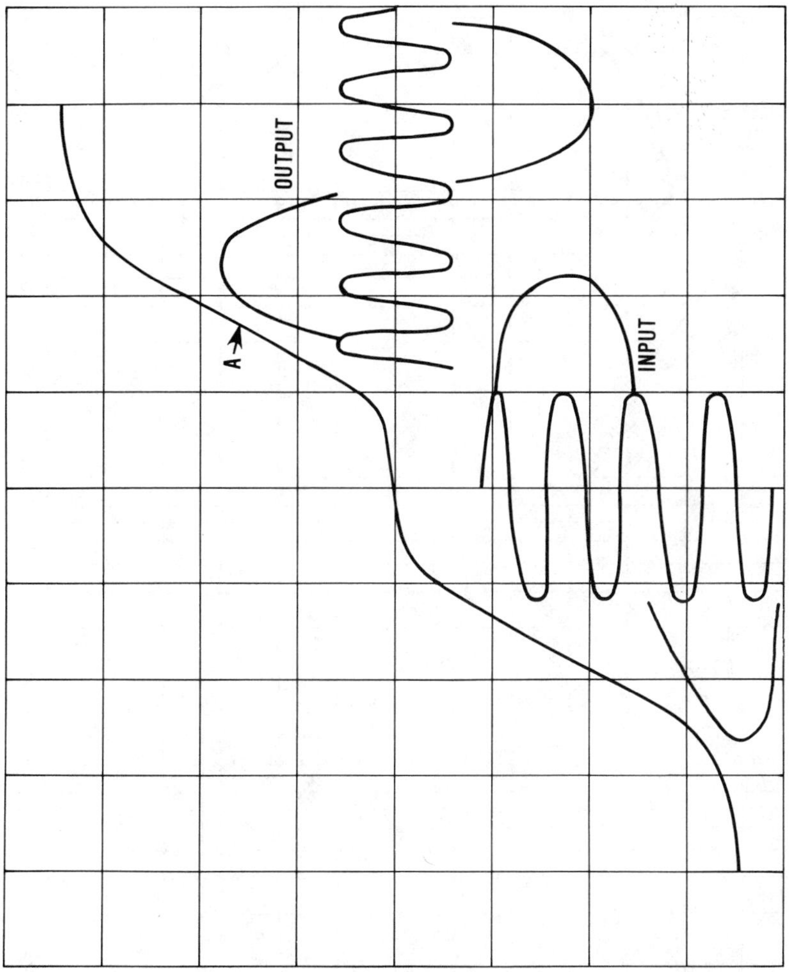

Figure 9-1

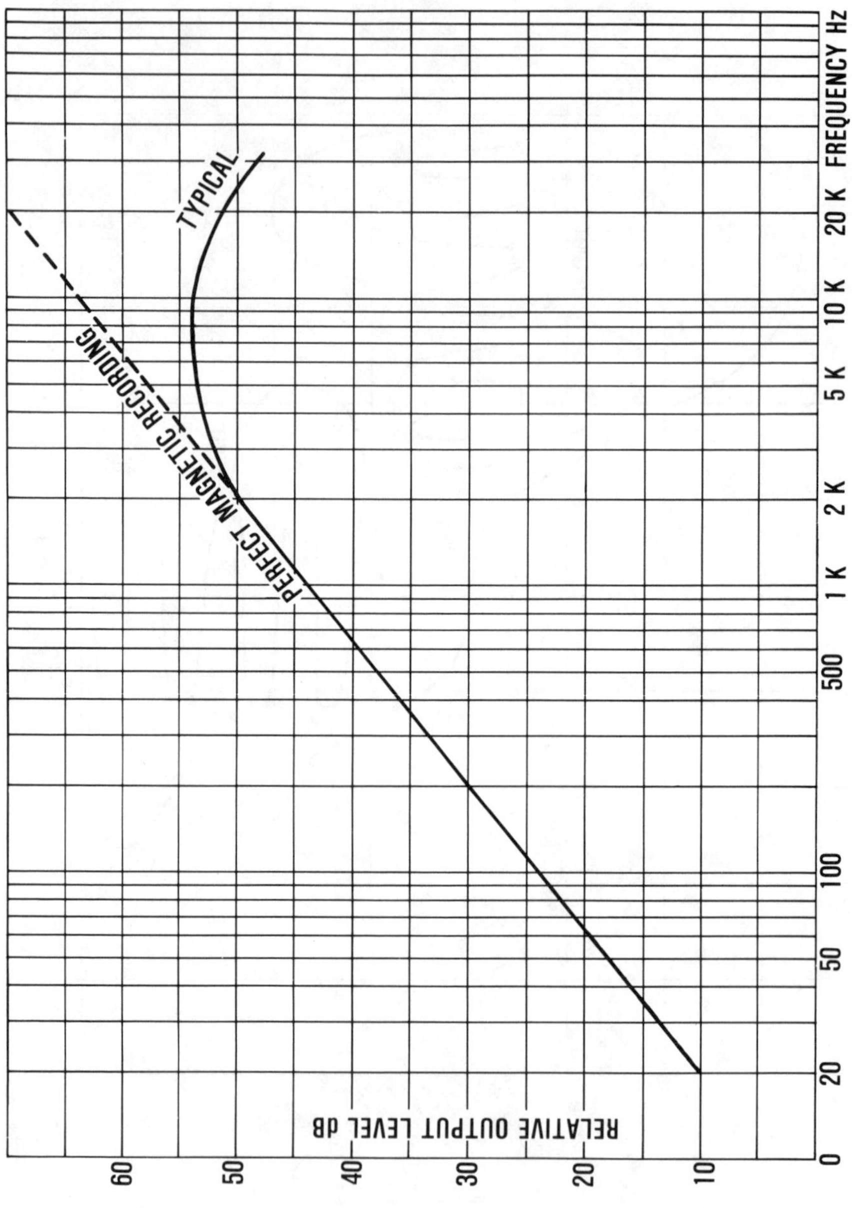
TYPICAL
PERFECT MAGNETIC RECORDING
FREQUENCY Hz
20 K
10 K
5 K
2 K
1 K
500
100
50
20
0
10
20
30
40
50
60
RELATIVE OUTPUT LEVEL dB

Figure 9-2

Open-Reel vs. Cassette Recording

Early magnetic recorders (1930-1960) were known as open reel. Plastic (polyester tape) tape coated with gamma oxide (Fe_2O_3) and ¼ inch wide by .002 inch thick was moved between two reels by a rotating capstan and over the record/play back/erase heads. The tape speeds in general use were 7½ and 3¾ inches per second. Open-reel recorders were developed to a high degree of performance. They still offer the best in all of the available attributes of good magnetic tape recording, short of the ultimate, namely, digital recording.

Many portable reel-to-reel tape recorders still are available. One of the disadvantages of reel-to-reel machines is the matter of changing reels which requires threading the tape through the drive mechanism and past the heads. This was simplified when packaged tape and machines to use it were introduced. Then came the cassette with its highly portable recorder/players and tape in a small slip-in package.

In order to get reasonable playing time, the speed was reduced to 1⅞ inch per second. With this arrangement, 281 feet of tape will record and play back for 30 minutes. By reversing the direction, you can extend the time to 60 minutes. The cassette tape is thinner and narrower than that used on open-reel machines. All of the changes made to provide cassette recording greatly increased the problems of making high quality music recordings.

The effort which went into the development of cassette recorders and cassette tapes resulted in a technology which gave us recordings of such high quality that they deserve to be classified as high fidelity. This fact, along with the popularity, availability and convenience of cassette recorders, makes cassette recording a suitable subject for treatment in this book.

Selecting a Fine Cassette Deck

The most complete listing of cassette decks can be found in Stereo Review's "Stereo Directory and Buying Guide." It lists the major characteristics and features in a factual manner, rather than giving a lot of opinions (which may or may not help you in your selection). In order to assist in your choice, take the following comments into consideration:

FRONT LOADING: Pretty much the standard and generally the most convenient.
THREE HEADS: Recommended for their convenience and value in monitoring tapes as they are being recorded.
DOLBY: A practical "must" to make recording and playing back compatible tapes possible.
STEREO: Taken for granted.
MOTORS: Several types acceptable. Electronic servo, electronic governor, separate capstan, and reel motors, etc.
METERS: Wide range VU or, better still, peak reading meters, plus LED level meters, at least indicating maximum normal level and overload.
BIAS: Three level Fe_2O_3 CrO_2 Ferrichrome, and metal, with some means for adjusting or calibrating. Best with test signals built-in.
EQUALIZATION: 70μs and 120μs minimum. (T = RC).
WOW AND FLUTTER: 0.6 or better.
SIGNAL-TO-NOISE RATIO: 60dB with Fe Cr tape, Dolby off.
FREQUENCY RESPONSE: 30-16,000Hz (Fe Cr) plus or minus 3dB or better.
CONTROLS: Input level controls (individually adjustable common control very convenient); output level control convenient but not indispensable; bias switch, 3 position and fine bias control; 2-3 position equalizer (70 μ s and 120μ s minimum); lever controls: play, record, rewind, fast forward, stop, pause and eject; speed adjustment; stereo/mono switch; Dolby.

JACKS: Microphone input (2) left and right; line input (2); ear phones (stereo); power cord.

Choosing a Compatible Tape for Your Deck

Recording high fidelity music on a cassette deck tape is pushing the state-of-the-art to its limit. No tape is perfect or able to record and reproduce exactly what is fed to the recorder input. Some tape characteristics can be measured quite accurately. Others can be heard or observed on a cathode ray oscilloscope and evaluated only qualitatively.

Frequency response, distortion, overload level and noise level can be measured for a given tape on a given recorder if one has the right measuring equipment. There is a catch, however; any measurements made apply strictly to only one tape, the one used in making the measurements. Tapes are made in batches. The coating can vary from batch to batch. The manufacturer may vary or even completely change his coating mix from batch to batch. The information provided by the manufacturer for a given tape is generally very meager. High grade, name brand tapes are less likely to suffer from these variables than other tapes.

Defects that are not easily evaluated quantitatively are primarily due to an imperfect recording surface. This may be aggravated by dust particles picked up by the tape. Listening to a recorded pure tone and/or observing the recorder output on a cathode ray oscilloscope will show up these defects. A 100Hz tone will show defects in the form of thumping sounds. A 1kHz tone will show clicks and amplitude fluctuations. A 10kHz tone will show skips and amplitude changes and fluctuations. If these effects are substantial, they will be readily heard in the reproduction of recorded music.

The choice of tape depends to a degree on the deck characteristics. In other words, the best reproduction of a given tape on a given deck may not insure the best results when using the same tape on another deck. This is due to three possible and important differences between decks. First, bias levels may be different, thereby operating the tape in a different part of its magnetization curve. Second, the equalization may be different although 70μs and 120μs should mean the same thing to all manufacturers. Third, the Dolby level may be set differently, thereby distorting the compression, expansion and frequency equalization. Decks with bias and Dolby level adjustments can be adjusted for optimum results with a given tape.

Tests were made with the three tone (100Hz, 1kHz, 10kHz) generator described in Chapter 6. This was done without bias adjustment or equalization in order to find the inherent frequency response of the tapes. Four chromium tapes were rated "excellent," having no more than 2dB difference between the 1kHz and the 10kHz responses. These four premium tapes are:

BASF Professional II
Maxell XL II
Memorex High Bias
TDK-SA

Eight tapes were rated "good":

BASF Professional I
FUJI FX-2
Maxell XL-1
Maxell UD
Memorex MRX3
TDK-D
TDK-AD
Realistic Super Gold

Testing Tape Surface Quality

Early magnetic tapes had a relatively rough surface. This showed up vividly when one recorded a steady tone and then played it back. The tone fluctuated, popped, wavered and otherwise signaled its rough and uneven surface. This problem was partly mechanical and partly due to the uneven dispersion of the magnetic particles in the oxide coating.

Magnetic coatings have been greatly improved over the years, and today some have attained a high degree of magnetic and mechanical perfection. The quick test of looking at the surface to see if it is glossy has some merit but is far from conclusive.

It was decided to attempt a more quantitative test. A setup was made as shown in the block diagram of Figure 9-3. A steady tone of 1kHz at -3dB was recorded for one to two minutes then played back, recording the peak level on a graphic recorder. The recording paper was moved at a rate of 3 inches per minute. The resulting graphs were very revealing.

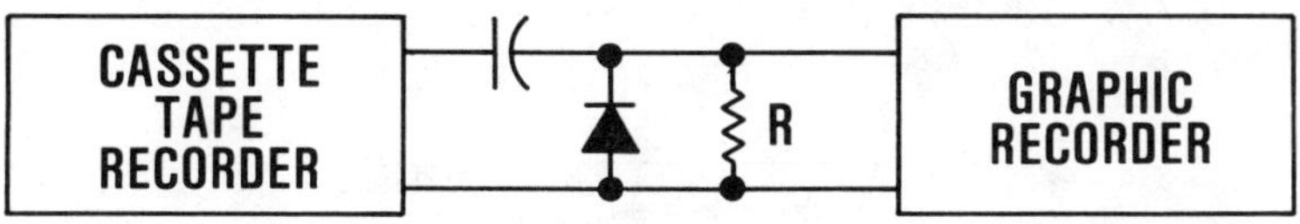

Figure 9-3

The highly rated tapes were noticeably smoother, to such a point that in some cases practically no wavering could be detected in the recording. The highest rated tapes are also the most expensive. But some fair and moderately priced tapes showed rather well. The cheap, low grade tapes, however, were all rough in response.

An unexpected result was that some previously used tapes showed serious bumps and clicks. This may be due to scratches and even creases and wrinkles.

Do not rely on many of the claims made on the tape package. Many very poor tapes are labeled in a misleading manner. One of the poorest tapes found in this test was labeled "High Density Hi Fi. High Energy. For Professional Use." Apparently, truth in advertising has failed in the cassette tape field, at least to a considerable extent. It is to be hoped that this industry will adopt more accurate and descriptive terms. Another problem in this field occurs when companies improve their coating formula or method without changing the description of the tape. Dating is suggested to help overcome this problem.

Figure 9-4 shows a number of recordings made to evaluate the surface and coating characteristics of various tapes as described above.

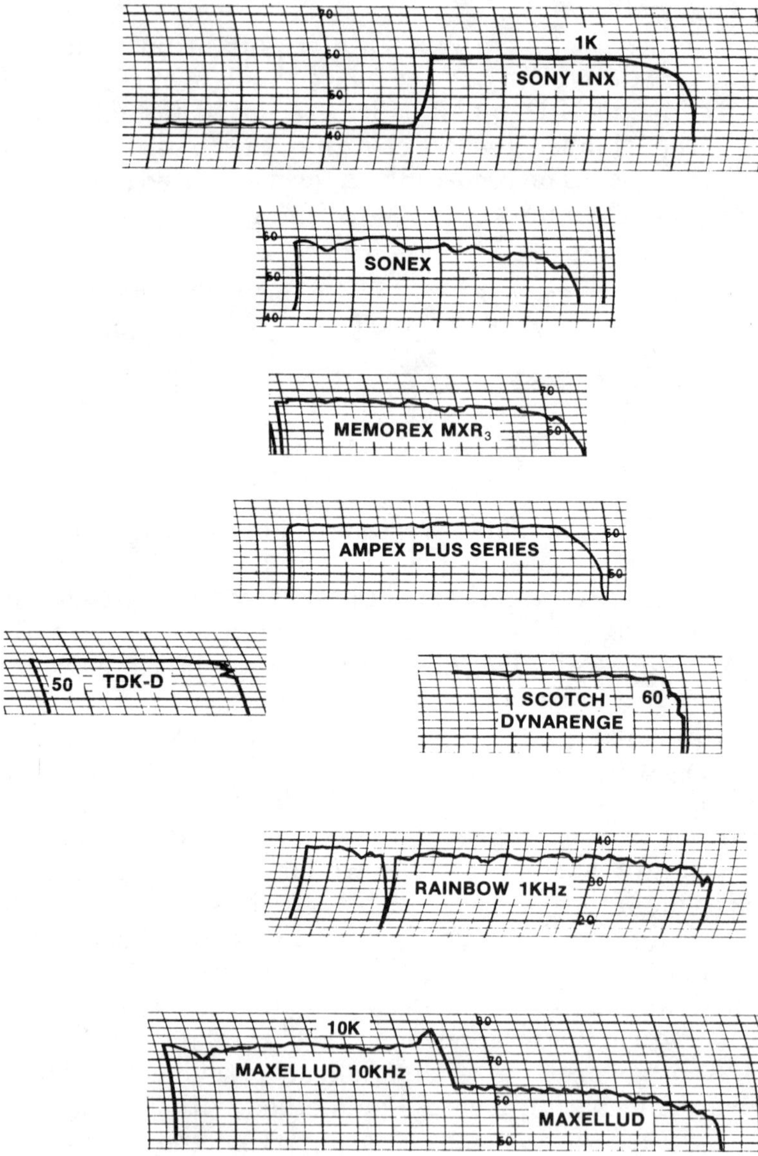
1K
SONY LNX
SONEX
MEMOREX MXR3
AMPEX PLUS SERIES
50
TDK-D
SCOTCH
DYNARENGE
60
RAINBOW 1KHz
10K
MAXELLUD 10KHz
MAXELLUD

Figure 9-4

Equalization of Magnetic Tape Recording

As previously mentioned, magnetic recording on magnetic tape has a highly nonlinear output versus frequency characteristic mainly due to two important factors. If a recording is made with a constant current to the recording head, it plays back with a rising characteristic versus frequency. This is due to the fact that the playback voltage generated by the playback head is proportioned to the rate of change of flux in the recorded pattern. This is the same as saying the playback voltage is proportional to frequency. Thus, in order to have a flat output versus frequency, an equalizing circuit must be used which attenuates the output linearly proportional to frequency. Figure 9-5 shows how the playback voltage rises with frequency and how the playback equalizer provides an overall flat response.

At the highest frequencies, there is a falloff due to characterstics of the tape, the recording and playback head gaps and the tape speed. One of the major problems met in tape characteristics is due to magnetic saturation. If one attempts to improve high frequency response when it is falling due to saturation, no improvement may be produced. The response will be improved by lowering the recording level, but then the question is whether or not one can hear the high frequency components at the reduced level. Raising the high frequency response after playback may help, but it may also increase noise.

It is in this high frequency region that the greatest difference exists between tapes. Ordinary tapes have poor high frequency response, regardless of how they are described. Premium tapes, particularly chromium dioxide tapes, have much better high frequency response; metal tapes have the best response, mainly due to higher saturation levels at high frequencies. See Figure 9-6.

If one assumes that the high frequency components are substantially lower in amplitude than the midrange, one can preemphasize the highs without saturating the tape. However, don't expect a great copy using inferior tape. Really good copies can be made only with the best equipment and on high grade tape. The recommended procedure is to use a three-head machine, monitoring the recording as it is made. At best, this is a time-consuming effort.

A practical question arises in connection with copy and that is with respect to Dolby compensation. A tape made with Dolby should be played back with Dolby deemphasis. An exception would be playing on an ordinary portable cassette record/player which may not reproduce highs very well. In any event it would be helpful to label tapes made with Dolby as such. (For a description of Dolby see page 199.)

Another practical question in connection with tape copying is recorder speed. Generally, speed accuracy of ±1 percent is sufficient. However, if one plays a tape at -1 percent on a recorder at +1 percent, a difference of 2 percent is produced in the copy. The best plan is to play the tape on the same machine on which it has been recorded with the assumption that it will play back at the same speed at which it was recorded, thereby cancelling any speed error.

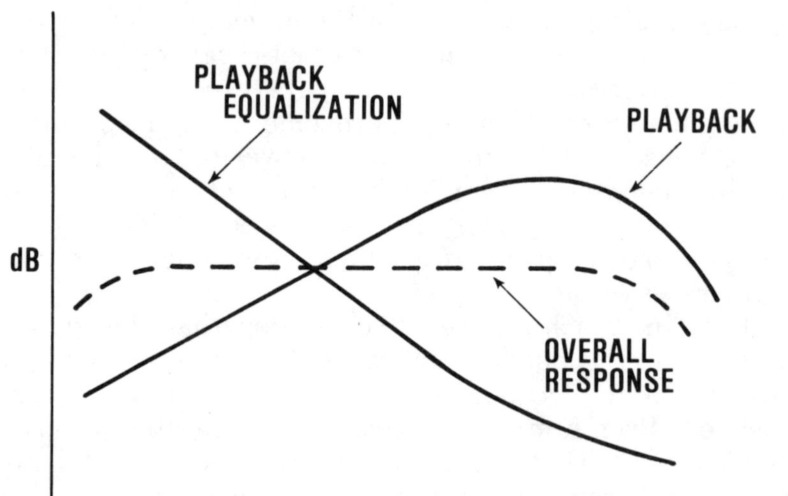

Figure 9-5

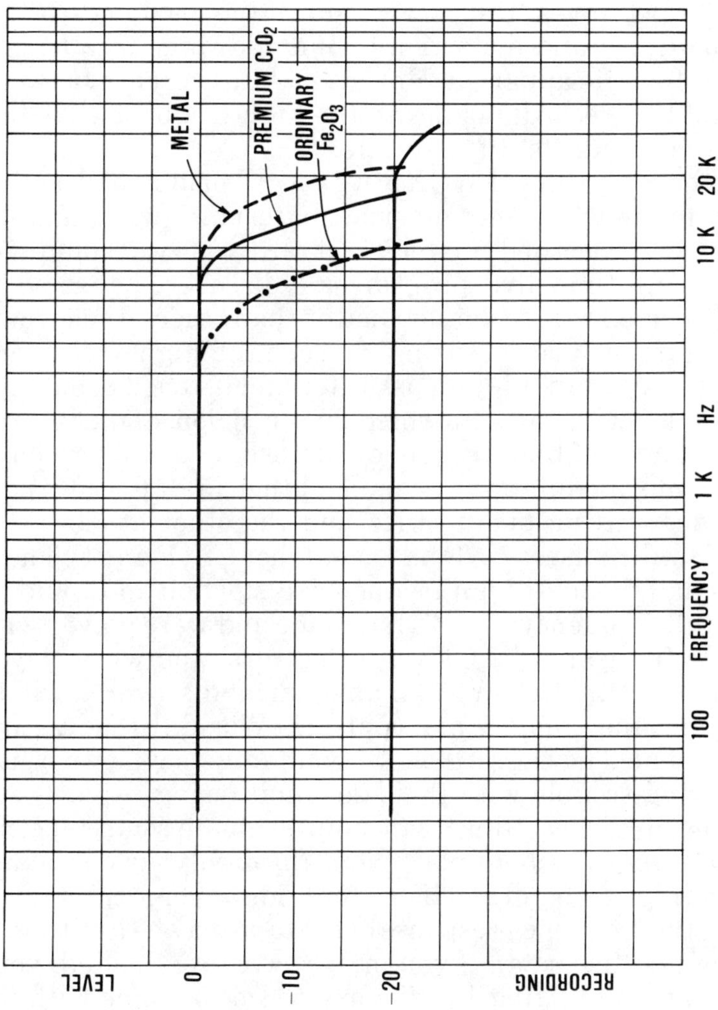
METAL
PREMIUM C_rO_2
ORDINARY Fe_2O_3
100
FREQUENCY
1 K
Hz
10 K
20 K
RECORDING
LEVEL
0
-10
-20

Figure 9-6

Evaluating Your Tapes

How can you tell how good the tape recordings you have made really are? A simple listening test should separate out the really poor ones. But how good are the good ones?

Muddiness and distortion in the midrange frequencies are usually quite apparent. However, the difference between good and bad recordings lies in the high frequencies which are more difficult to differentiate.

In the early days of WQXR we used World Broadcasting records for much of our musical material. I was unhappy about using recorded material at all, but we could not generate full-time live programs. I discovered that some records seemed to sound much better than others. I suspected that distortion was creating discordant high frequency components. I built a high-pass filter and listened to only the high frequencies. This confirmed my suspicion. Some records contained a lot of high frequency discord. Probably much of this was intermodulation distortion. I then built two low-pass filters, one cutting off at 5kHz and the other at 3.5kHz. I listened to all of the records and rated them, A, B and C. The A records had the least high frequency distortion and could be played full frequency; the B's had some and were played with the 5kHz filter and the C's were the worst and were played with the 3.5kHz filter. When management discovered what I was doing they made me confront World Broadcasting management. They agreed with my findings, saying they had been having trouble with their gold sputtering.

This same procedure can help evaluate your tape recordings. One way is to play them through your equalizer. Raise the highest band and listen for highs to come through. If none do, the frequency response of the tape may be limiting. If they sound discordant, you may have overloaded your recorder, thus creating IM distortion. Do the same with the next highest band. Deep bass can also be examined by raising the lowest band level.

Listening Techniques

Many sources recommend listening to make a final judgment on such components as loudspeakers, phonograph cartridges, and so on. The question is: How does one listen? What does one listen for? There are a number of keys to the technique of listening.

1. Listening to the human voice. Listen for naturalness and easy intelligibility. Reject muffled sounding voices; boomy voices; scratchy voices; voices in a barrel; sharp, but not clear, female voices. Ideally, a human voice should sound so real that you have to look to see that he or she is not actually there in person. It can be done and I've experienced it.

2. Listening to music. Sometimes just listening to music for a prolonged period, say, five or six hours, can reveal distortion and/or unnatural coloration of the sound. If the music is still very pleasant and you do not have any fatigue or annoyance, the system is, at least, rather good. This holds true only if you are not so used to listening to the system that you have developed a tolerance for its particular brand of distortion and/or coloration.

Listen to familiar music. Listen to piano music. Does it give the illusion that the piano is real and in your listening room? If you are a musician familiar with pianos, you should be able to identify the maker.

Organ music provides a good test source. If the deep bass notes are full and clear and do not break up, you have made a good test of the low frequency response of your system. A recommended record for the purpose is Crystal Clear Records' Virgil Fox, "The Fox Touch," No. CCS-7001.

Music involving a number of instruments is another source of test material. Listen for each instrument. Can you hear each one clearly? Various instruments show the response of your system in various ranges. Some, such as cymbals and bells, stress the transient and high frequency response. Clear definition shows lack of serious distortion. Loud passages test the undistorted power handling capacity.

3. White noise. Tuning between stations in the FM band provides a quasi white noise source. It should sound very smooth, without any impression of high or low pitch emphasis. It requires some familiarity and experience to make a white noise test that is meaningful.

4. Stereo. Stereo should give the illusion of depth. Instruments should be identifiable as to their actual position in the music source. Placement and balance between speakers is of utmost importance. Find the area where stereo listening is well defined. More on this subject will be found in the section entitled, "The Listening Room Is Important to Hi Fi" in Chapter 10.

Bias and Bias Adjustment

When a magnetizing force is applied to a magnetic material it responds in accordance with its characteristic magnetization curve as shown in Figure 9-1. The response starts slowly for very small magnetizing flux, increases and becomes almost linear over a region, and then decreases until it becomes flat at saturation. In order to create low distortion magnetic recordings, the response must be linear. In some early magnetic recorders, a DC bias was used to place the operation in the center of the linear portion of the curve (point A in Figure 9-1). This was found to be a critical factor and differed widely among tapes. It was discarded in favor of AC bias.

AC bias was a big improvement and is the generally accepted method of making good magnetic recordings. AC bias is not easy to explain nor understand, but how to use it is quite straightforward. The AC bias is a high frequency signal added to the signal to be recorded. This bias signal should be well above any audio frequency to be recorded and is generally on the order of 70kHz. In effect, it offsets the signal being recorded in both positive and negative directions, thereby causing the signal being recorded to skip the toe of the magnetization curve and to fall in the linear parts as shown in Figure 9-1. On playback, the recorded signal is recovered from the linear areas and the high frequency bias is eliminated by the response of the playback circuit.

The magnetization curves of magnetic materials used in magnetic recording vary over a wide range. The softer ferric oxides are much more easily magnetized than the chromium dioxides, for example. This means that a bias suitable for ferric oxide will be too low for chromium dioxide and vice versa, and there are formulations falling between the two that make things more complicated. Improper bias amplitude causes the recording to take place above or below the ideal linear range on the tape.

Figure 9-7 illustrates the distortion produced by using too low a bias on a CrO_2 tape while Figure 9-8 illustrates the distortion produced by using too high a bias on a ferric oxide tape. Figure 9-9 is an actual oscillograph of a 1kHz signal recorded and played back from a ferric oxide tape while using a chromium dioxide (high) bias.

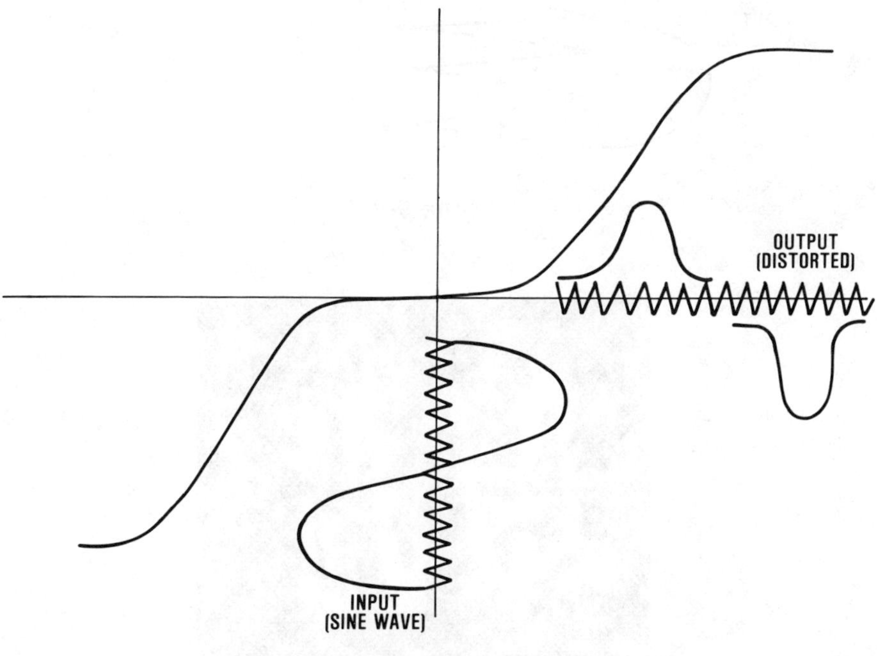

Figure 9-7

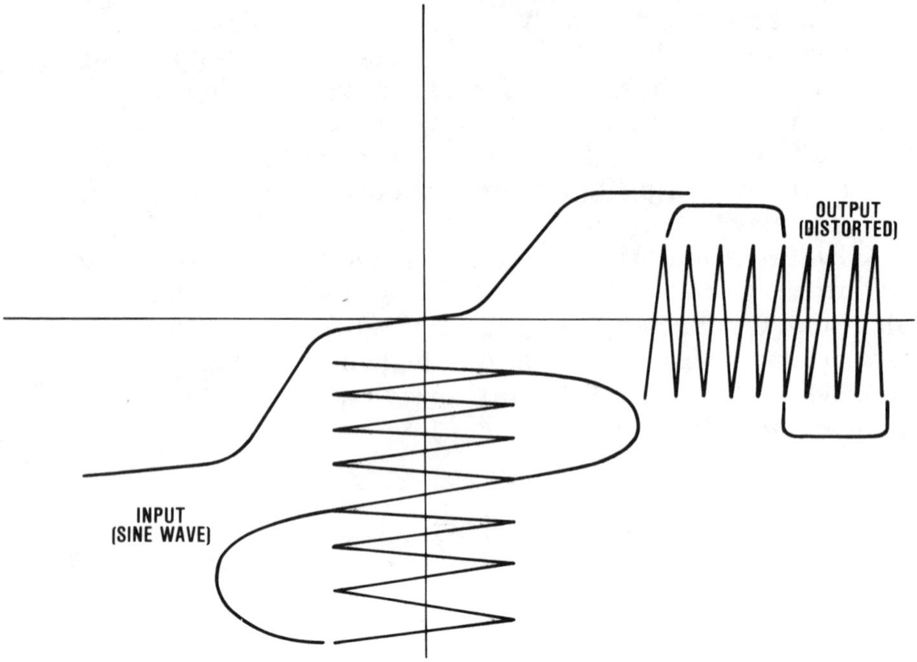

Figure 9-8

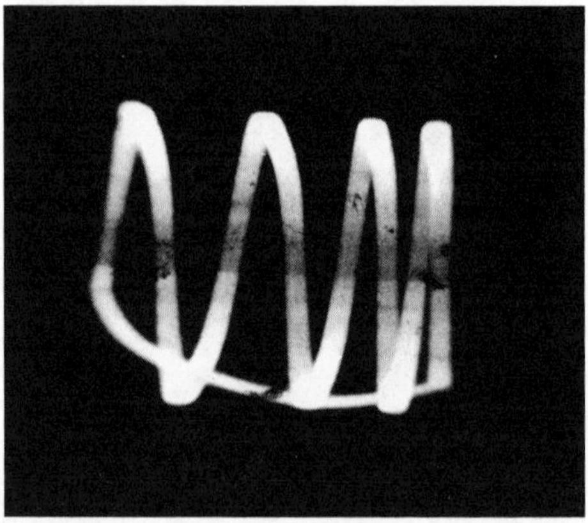

Figure 9-9

Simple, portable cassette recorders have a single fixed amplitude bias suitable for the standard ferric oxide tapes. The standard cassette deck is generally equipped with two switch-selectable bias levels, one designated standard and the other, CrO_2. The manufacturers of these decks set a low bias level for ferric oxide and a high bias level for chromium dioxide using two selected tapes as standards. For example, TDK-D may be used for the ferric oxide and TDK-SA for the chromium dioxide on the assumption that these tapes represent state-of-the-art oxide coatings in the two classes.

However, the magnetic tape art has proliferated and many variations are now available. Most tapes specify low or high bias but that does not mean that they are optimum. The more sophisticated tape decks have adjustable biases. By calibrating these adjustments and testing each tape to be used, you can get much closer to the ideal. The adjustments should be made in accordance with the operation instructions for each particular machine.

Metal tapes provide increased midrange headroom (about 4dB) and a slight extension in the high frequency range, but at a substantial increase in cost over ferrochrome or chrome tapes. If you have a deck with metal tape bias provision, you may also have material benefiting from this greater dynamic range. The essential advantage provided by metal tape is that its saturation level is higher than in other tapes and that it has a slight advantage at the high frequency end of the spectrum. These advantages show up most clearly when there is a lot of high frequency energy in the material being recorded and the recording is being done at a higher than usual level.

Dolby Noise Reducers Not All the Same

Dolby is a patented system for reducing noise in magnetic recordings. The basic Dolby system splits the audio spectrum into four parts, expands, contracts, equalizes and so on to provide a 10dB gain in signal-to-noise ratio. Tapes recorded with Dolby must be played back with Dolby in order to obtain the advantages of the system and to have the proper balance in the playback. Playing a Dolby recorded tape on a non-Dolby machine or even on a Dolby machine of another manufacture may result in a degraded result. Also there are similar noise reducing systems which are not compatible with Dolby. The best procedure in any case is to play back a tape on the same machine on which it was recorded and with the same compensating setting and equalization.

Ideally, Dolby must relate to the particular tape being used. Some machines provide for making this Dolby calibration.

Speed Adjustment Assists Copying

Speed accuracy within ±1 percent is generally regarded as sufficiently accurate for most recording and playback of magnetic tapes. However, a problem may arise when a tape is made on one machine, copied on another and played on a third. There can be a cumulative error, in which case 1 percent can escalate to 3 percent, thus becoming objectionable.

Some decks have speed adjustment means. In order to make a speed adjustment you must have a tape with an accurately recorded tone of known frequency. You also require a source of accurately known frequency which can be adjusted to the tone on the tape. The tone on the tape and the reference tone are then compared by listening to the two at the same time, or by displaying them on an oscilloscope. The speed adjustment is then varied until the tone from the tape exactly matches the reference tone.

A simple way to determine whether or not the speed of a second deck is equal or close to the speed of a first deck is to record the note from a pitchpipe on the first deck and play it back on the second, while listening to the actual tone from the pitchpipe.

Care of Your Deck for Long Life

Magnetic tape must have a very smooth and clean surface for good reproduction of music. While the tape magnetic coating is very tough, it does wear and pick up dust. If surface particles or dust accumulate on the record/playback head, it can seriously interfere with good reproduction. Also the record/playback head can become magnetized, causing distortion of recorded music.

The cover over the cassette compartment should be kept closed. The record/playback head, as well as the erase head should be cleaned and demagnetized on a regular basis. Cleaning and demagnetizing should be carried out in accordance with instructions contained in your owner's manual.

Tape Counters: What Do They Really Mean?

The implication is that a tape counter measures tape as it passes through the tape recorder or player. Actually, it indicates the rotation of one of the tape reels. On some machines the supply reel rotation is indicated while on the others it is the take-up reel. Furthermore, the ratio of reel rotation to indication is not standard. Thus, a metered point on one machine is quite different from that on another. The indications are roughly twice as fast at one end of the tape as they are at the other end. This again depends on the diameter of the reel hub.

All this means is that for a given tape recorded or played on a given machine, the indicator will indicate a given point on the tape. This indication may not be exact, however. The accuracy may depend on whether the tape has been completely played on the same machine, whether it has been fast rewound, or the tension on the tape has been changed in some other way as it is wound and unwound from the reel carrying the counter mechanism.

In other words, a tape counter is an arbitrary indicator of tape position on a given machine for a given tape. Also it assumes resetting to zero at the start of the tape each time it is used.

Tape Surface Testing with IM Meter

The IM Distortion Meter indicates low frequency modulation of a high frequency tone. It thus provides means for measuring the low frequency modulating effect of rough tape surface on a recorded high frequency tone. Instead of recording the low frequency signal along with the high frequency, it is merely necessary to record the high frequency tone alone. The tape surface roughness will do the modulating.

A roughness measurement is made in the same way an IM distortion measurement is made, except that only the 5kHz tone is recorded (no 60Hz component). When played back, the tape surface roughness reads out as a percent IM.

I measured a number of tapes in this manner and found percentages of roughness modulation ranging from about 1 to 5 percent. There was a rather unsteady average percentage with sudden peaks of about double the average. Looking at the signal after recording, and both before and after demodulation, confirmed these results. The higher percentages were found on the cheap tapes and the lower figures on the higher grades. These results would seem to indicate that in spite of highly developed cassette recorders, a substantial source of disturbances resides in the tapes no matter how highly refined.

Visualizing the Roughness of Tape Surface

A somewhat more visual method of checking tape roughness is to record a steady tone, play it back and photograph the result on a cathode ray tube. This was done and the results are shown in Figures 9-10 and 9-11 for a cheap tape and a premium tape, respectively. You can see that while the premium tape is definitely much smoother than the inexpensive tape, it still is not perfect. The irregularities are quite audible in the case of the steady tone on the inferior tape and all but inaudible on the premium tape. While it is a factor distinguishing poor from superior tape, it is less important than the frequency response in the two cases.

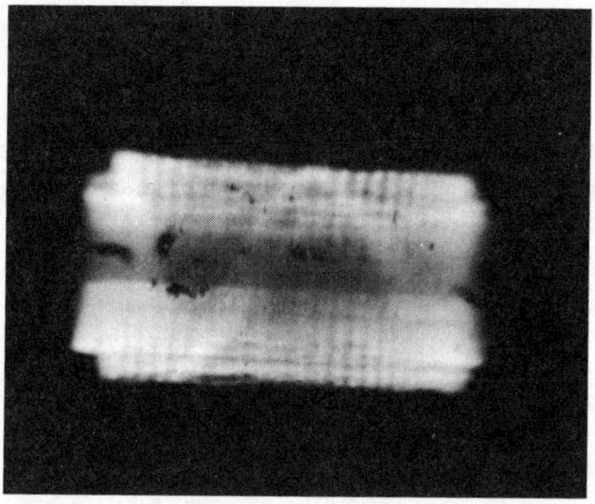

Figure 9-10

Figure 9-11

10

Key Factors That Contribute to Higher Fidelity

Listening Tests in General

The electrical signals generated by a microphone are amplified, used to modulate a transmitter (or recorded), transmitted, received, amplified, detected, amplified again and applied to the terminals of a loudspeaker. This entire process can be rigidly controlled and distortion and other effects closely predicted. However, what happens to the sound sources before they reach the microphone, and what happens to the speaker currents before they reach the listener's ears are complex processes and not easily controlled effects.

At the source end, the acoustical properties of the music room, auditorium, theater, outdoor band shell or other area, combined with the placement of microphones, determine the phasing, reverberation, echoes and other effects on the source sound before it is translated into electrical signals. The microphones have a rather secondary effect and mainly affecting low and high frequency response, although their directional qualities also modify the music room effects.

At the other end, there is another set of conditions affecting the result that reaches the listener's ears. First, the loudspeaker does not faithfully convert the applied audio current to corresponding sound pressure. Perhaps more importantly, the listening room emphasizes some frequencies, deemphasizes others, reverberates and reflects. A loudspeaker that tests extremely well in an anechoic chamber, can be substantially modified in its apparent response in an actual listening room.

Because it is not easy or even possible to predict the performance of a loudspeaker from measurements sometimes made in an anechoic chamber, listening tests are widely advised. To be realistic, of course, the listening tests should be made in the room where the loudspeakers are to be used. This is most difficult if you must compare several loudspeakers. A compromise is to listen in a hi fi dealer's listening or demonstration room. The question then is, how closely does such a listening test resemble your final listening room?

The Listening Room is Important to Hi Fi

Listening rooms should be what one would call a "quiet room." This means several things. It should be a room as free from outside disturbances as possible. It should be a room of moderate size, so that a comfortable distance can be maintained between listener and speaker, especially when stereo is involved and/or there are a number of listeners. The room should be acoustically quieting, that is, it should not be too live—a too live room sounds noisy. Nor should it be too dead—a too dead room gives a hushed feeling. Hard reflecting and opposed surfaces should be avoided. The room should be large enough so that speakers can be moved about for best results and so that they are not confined to corners or close to flat walls.

A great deal of time, thought and experimentation have gone into determining the optimum reverberation time of the ideal listening room. Reverberation time is defined as the time required for a sound in a given room to fall from its original level to a level 60dB down. Unfortunately, there is no ideal reverberation time under all conditions. Speech, for example, is most intelligible in a room having a reverberation time of around 0.75 second. On the other hand, a large orchestra in a large hall should experience a reverberation time of two to three seconds. Thus, a listening room will have an optimum reverberation time which depends on its size and on the type of material being reproduced. This will vary from about 0.6 second for a very small room and fast speech reproduction to two seconds or more for an organ playing slow music in a large room. By small room is meant one of about 1000 cubic feet and a large room up to about 250,000 cubic feet.

A Closer Look at a Moderately Sized Room

We will consider in more detail a typical, comfortably large room. The experimenter can then extrapolate to suit his or her own particular case. A room $20 \times 15 \times 8$ feet will have a cubic volume of 2400 cubic feet. In such a room, a one second reverberation time will fall about midway between the optimum for fast speech and slow organ.

Our $20 \times 15 \times 8$ foot room has a volume of 2400 cubic feet. Now the reverberation time T is defined as follows: $T = V \times 0.049/AS$ where V is the volume of the room in cubic feet. A is the average absorption coefficient and S is the total surface area in square feet. Thus, $AS = A\ (20 \times 2 \times 8 + 15 \times 2 \times 8 + 20 \times 15 \times 2) = A \times 1160$. Returning to the equation for reverberation time, if T is equal to 1 second, solving for 1160A, the total absorption of the room, we get AS (total absorption) is $2400 \times 0.049/1 = 117.6$. In order to find the total absorption, we must add all absorbing areas and materials.

Many areas contribute very little to sound absorption and can be disregarded. Plaster walls are one example. Following are some of the more important materials and their coefficients of absorption. Multiply the coefficients by the square feet of the particular material. When you reach the total of 117.6, you have approximated the ideal listening room acoustics.

Carpet	0.2 (no pad)
Carpet	0.4 (with pad)
Draperies	0.1 (lightweight)
Draperies	0.5 (very heavy)

The following items are absorption per unit and add directly to the total:

Upholstered chair	0.5
Sofa	1.5
Audience (per person)	0.5

If your room is substantially different, you can recompute the reverberation time using the above formula.

Experimenting with Speaker Locations

Speaker placement in a listening room is a matter that depends on many factors. These factors include: room size, room acoustics, room configuration, speaker size, speaker characteristics, type of material to be listened to, number of listeners; mono or stereo system, power available, etc.

The speaker or speakers should be unobstructed. There should be no furniture or other objects which could obstruct or deflect sound between the speakers and the listeners. The speakers should not face a major object which could deflect the sound in an unwanted direction. This means that in a good listening room the space between speakers and listeners should be as free and open as possible.

Consider Corner Speakers for Boosting Bass Response

When a speaker is placed in the corner of a room, the walls and floor create a horn effect which may substantially increase the effective bass response of the speaker. If corner speakers are to be used for stereo, two identical corners not more than about 12 feet apart are required. The enhancement of the bass may require tone control or equalization to cut bass for a smooth result, unless one enjoys over heavy bass. Raising the speakers from the floor two or three feet will reduce the bass reenforcement somewhat.

A smaller speaker having a deficiency in its bass response may be substantially improved by using a corner location. However, if the bass just isn't there, no manner of corner placement will restore it.

This placement of speakers assumes that the walls of the corner are bare and therefore good reflectors. If they are padded or heavily draped, the bass sound will be more or less absorbed, and little bass boost will be produced.

Influence of Walls on Woofer

When a speaker is placed close to a reflecting wall, the wall and floor act to reenforce the woofer at very low frequencies and cause cancellation at some higher frequencies as does the corner placement of speakers, but to a lesser degree. Moving the speaker out from the wall and raising it from the floor minimizes these effects. The distance you must move and raise the speakers depends on the characteristics of the speaker and the wall, and the listener's preference. At distances greater than three feet, the effects practically disappear. A heavy carpet, of course, will make raising the speaker less important, although the damping effect of the carpet decreases as the frequency is lowered.

Reverberation in a Room Can Be Too Much or Too Little

Acoustic studies of reverberation in rooms and its effects have been going on since Sabine did his classic studies at Harvard University in the early 1900s. Up until that time, auditoriums had been built with varying acoustical properties and little scientific basis. Sabine pioneered a thorough study of sound in rooms, and the effects of reverberation on articulation and musical quality. His book *Collected Papers on Acoustics*, published in 1923, brings together the far-reaching results of his studies.

Some of the subjects studied and reported by Sabine were: loudness; distortion of complex sounds by interference and resonance; confusion by reverberation, echo and extraneous sounds. He related the absorbing power of walls and other surfaces to room size. He pointed out that reverberation was a purely physical matter and so a subject for scientific investigation, but the question of what conditions are best for the distinct audition of speech or for the enjoyment of music in its various forms are matters of judgment and taste.

Subjective tests showed that the optimum reverberation time for one size room was not optimum for another size. The larger the room, the longer the preferred reverberation time. The optimum reverberation time for the best articulation was substantially less than that preferred for music, particularly certain types of music. Organ music, for example, required a longer reverberation time than did piano. To make matters even more complicated, reverberation time depends on the frequency of a tone being examined.

Definition of Reverberation Time. Reverberation time is defined as, the time a sound of a given pitch and intensity decays 60dB. Initially, it was taken as the time a sound decayed from some arbitrary level to inaudibility. Inaudibility is defined as 0dB. To make some usable standard, the definition assumes from 60dB to 0dB, hence, the decay of 60dB.

From Auditorium to Listening Room Affects Realism. Sabine's studies were made before the days of radio and hi fi. He was mainly concerned with auditoriums or rooms where speech and music were first-hand sources. However, many of his findings have meaning today, although a great deal more research has been conducted over the years.

Speaker Placement for True Stereo Effects

Placement of speakers becomes more critical when stereo listening is involved. Not only do walls and corners and listening room acoustics become more noticeable, but the distance from the listener to each speaker becomes important. It is assumed that stereo records and broadcasting is done with microphones spaced sufficiently far apart so that real stereo effects will be produced. In listening as well, the speakers should be well spaced. They should be spaced so that the listener is aware of sound coming from two sources, the two stereo speakers. The room should not be so reverberant as to intermix sound from the two speakers, thus making the two sources not well defined.

As an example, in a fair sized, not too reverberant room, speakers may be spaced about ten feet apart. The listener should be on a midline between the speakers and about ten feet away. Obviously, the two speakers should be identical in response or false stereo effects will be produced.

While there may be only one ideal spot for stereo listening with a given speaker arrangement, the further apart the speakers are and the larger the room, the more area is available for satisfactory listening. Since no exact formula has been established, this again is an opportunity for experimentation.

It goes without saying that stereo speakers must be operated in phase. Phasing can be easily checked by standing midway between the speakers and moving slightly one way and the other. If a dead region is encountered, the speakers are out of phase.

Bookshelf Speaker Placement for Optimum Listening. Bookshelf speakers are designed for what their name implies—placing them on a shelf. Since such speakers seldom provide much bass, mounting them near a wall has little effect. Of course, for stereo they should be well spaced apart.

It is not likely that a portable radio with speakers only a foot or so apart can render any realistic stereo effect except, perhaps, at very close range.

The Ideal Listening Room—How Close Can You Come to It?

When one speaks of the "ideal listening room," what is he or she really trying to say? Ideal for what? Ideal for whom? Ideal for what equipment? Ideal for what type of music or speech? About the best you can do is to try to describe a good all-around listening room and then try to arrange your particular listening room to best simulate this ideal room.

Figure 10-1 is a simple diagram of one hypothetical listening room. There is room to place the speakers about 10 feet apart and to listen at a position about 9 feet equidistant from the two speakers. The speakers are spaced from the wall and corners to avoid unbalance and overemphasis on the bass. For true stereo balance there is only one ideal listening position, which is as shown. If several people wish to listen at the same time, thy must sit very close together or use stereo headphones.

If corner speakers are to be used, the room can be rearranged in the other dimension with the speakers in each corner of the 12-foot wall. If bookshelf speakers are to be used, they should be raised from the floor. The midrange and tweeter speakers should be as nearly at ear level as possible so that the stereo effect is enhanced by maximizing the direct sound.

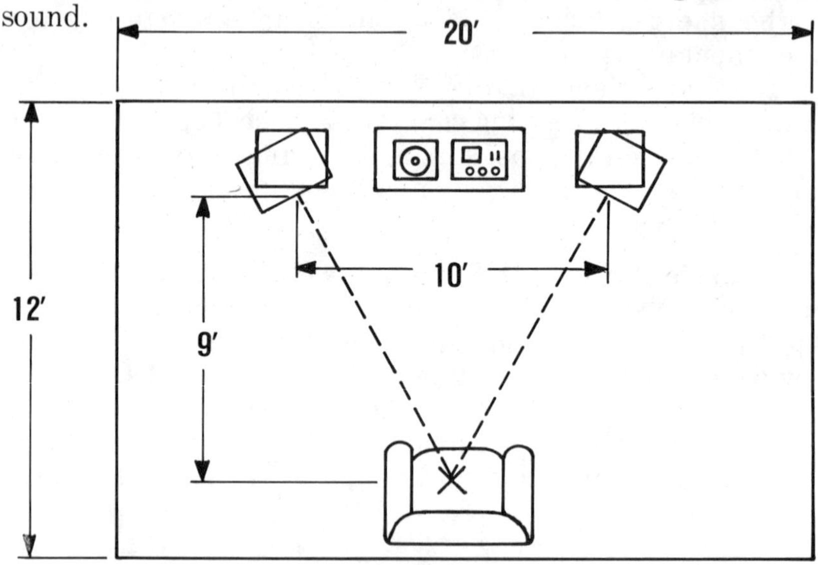

Figure 10-1

A listening room should not be too dead or too live. A decay time between one and two seconds is recommended. This generally calls for a medium heavy carpet covering most of the floor, some draperies and some padded furniture. Furniture, especially bookcases, is helpful in breaking up possible echo patterns. Here again, experimenting is useful in adapting your room to your system and listening tastes.

The demonstration rooms where high fidelity speakers and other equipment are displayed are generally on the dead side. You can expect more brilliance in your own average good listening room. Another difference between the demonstration and your use of the equipment may be that the demonstrator is using equalizers. You should check this possibility very carefully. On the other hand, if you have an equalizer, you can then create your own emphasis.

Equalizers have many uses. One is the correction of listening room inadequacies or other problems. Adding high frequency emphasis can increase the brilliance of the reproduced sound. Depressing midrange can correct for room resonance and excessive reflections. Adding bass can increase low frequency depth and resonance. Decreasing low frequency response can correct for low frequency room resonance or adapt boomy music more to your taste.

Equalizers are also useful for equalizing audibility when listening at low levels. Figure 10-2 is a series of equal loudness curves for various relative sound levels. For example, at the threshold of hearing, 0dB at 1kHz, it takes a level of 50dB at 50Hz to just be audible. Again a tone of 1kHz at 20dB level sounds as loud as a 50Hz tone at 64dB and a 10kHz tone at 30dB. As the level is increased, the low frequency audibility flattens markedly but the high frequencies show much less change. Thus, to make listening at a low level of 20dB sound natural, the very low frequencies must be boosted about 45dB.

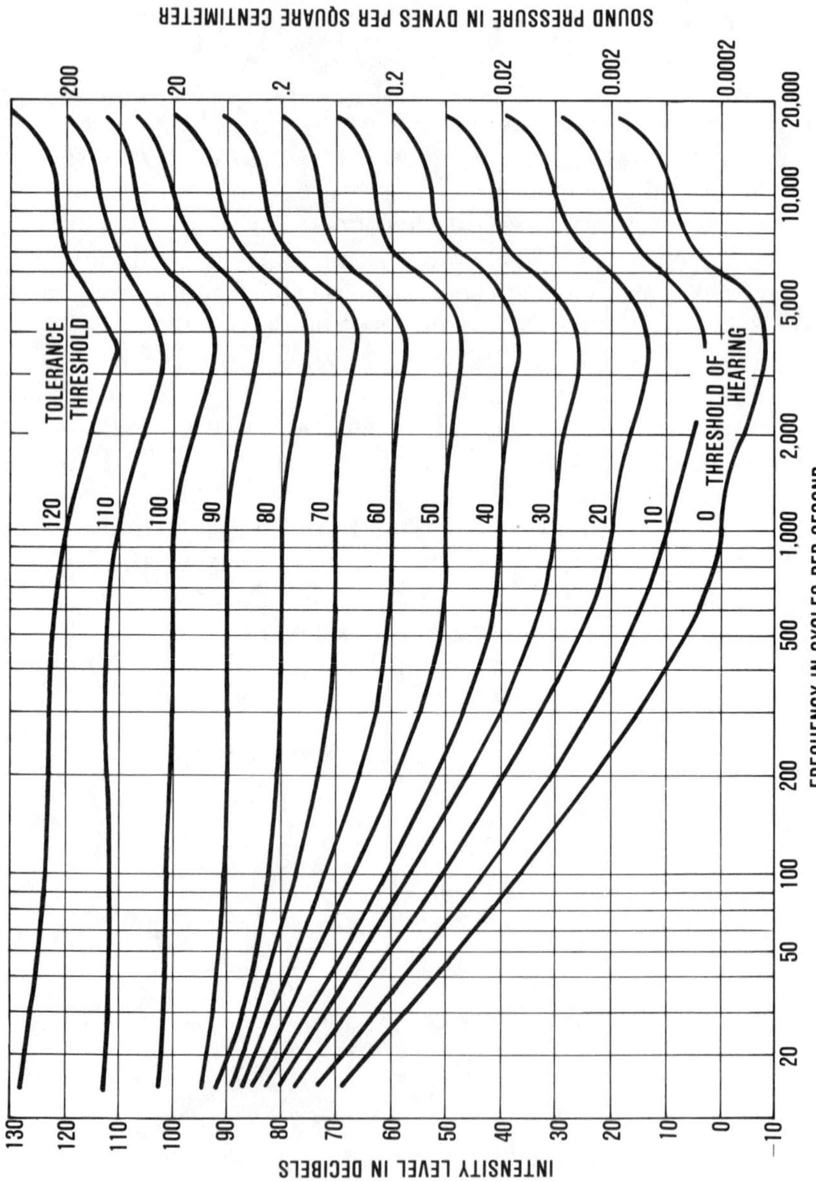
SOUND PRESSURE IN DYNES PER SQUARE CENTIMETER
200
20
2
0.2
0.02
0.002
0.0002
TOLERANCE THRESHOLD
THRESHOLD OF HEARING
120
110
100
90
80
70
60
50
40
30
20
10
0
INTENSITY LEVEL IN DECIBELS
130
120
110
100
90
80
70
60
50
40
30
20
10
0
−10
FREQUENCY IN CYCLES PER SECOND
20
50
100
200
500
1,000
2,000
5,000
10,000
20,000

Figure 10-2

11

Understanding Square Wave Test Results

Square wave testing is not only easy to perform but it is very revealing. A 500Hz square wave tells a great deal about the response of a system from a low frequency of 50Hz or less to a high frequency of 5kHz or higher. Even 10 to 20kHz response can be evaluated through careful observation.

In order to be able to interpret square wave responses, Figures 11-1 through 11-6 illustrate the most basic square wave repsonses that can be described as follows:

Figure 11-1 is an accurately represented square wave. This is the wave form produced by the 500Hz square wave generator described in Chapter 6. When observed on a cathode ray oscilloscope, the spot moves from left to right as indicated by the arrows. The vertical rise indicates the high frequency response of a given device and the horizontal plateau, the low frequency response. If the high frequency response is poor the vertical rise slopes to the right and the upper corner becomes rounded, as shown in Figure 11-2. If the low frequency response is deficient, the horizontal line tilts downward as shown in Figure 11-4. If the highs are accentuated, the upper corner peaks as shown in Figure 11-3. If the system has an inductive or mechanical oscillation due to lack of critical damping, a ringing is exhibited as in Figure 11-5. When the frequency response is limited to responses below the period of the square wave, slewing takes place as shown in Figure 11-6. It is not likely that your system or components will have such a poor response as to observe this slewing at 500Hz. It is more likely to show up at 10-20kHz. Incidentally, this slewing is the source of what is called transient intermodulation distortion or TIM.

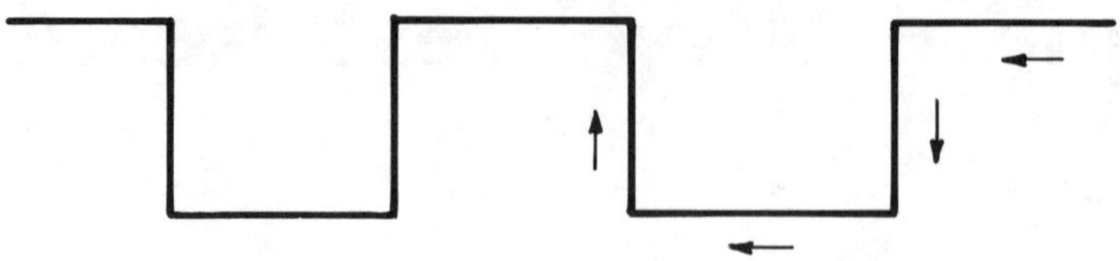

Figure 11-1

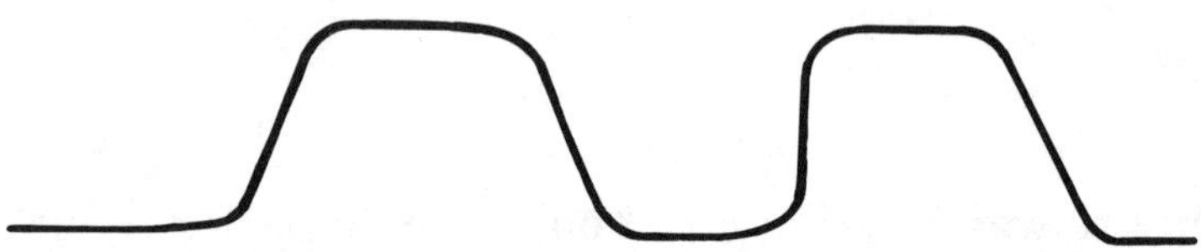

Figure 11-2

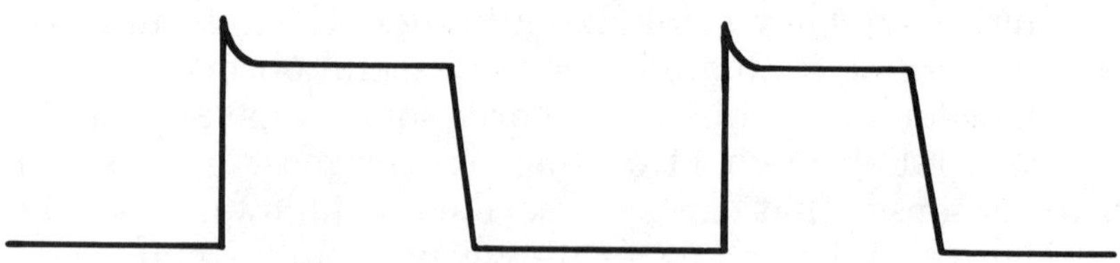

Figure 11-3

Figure 11-4

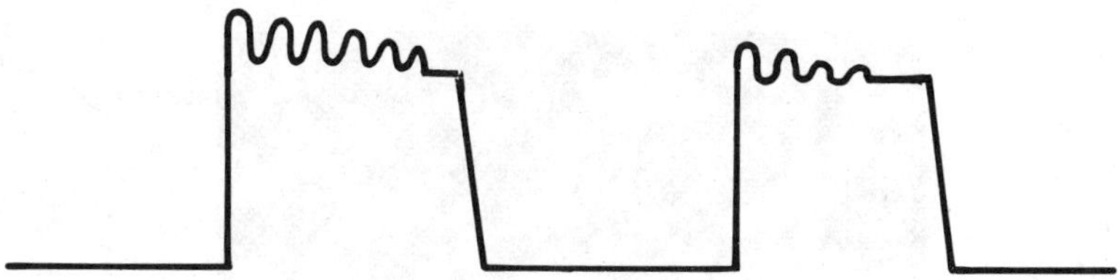

Figure 11-5

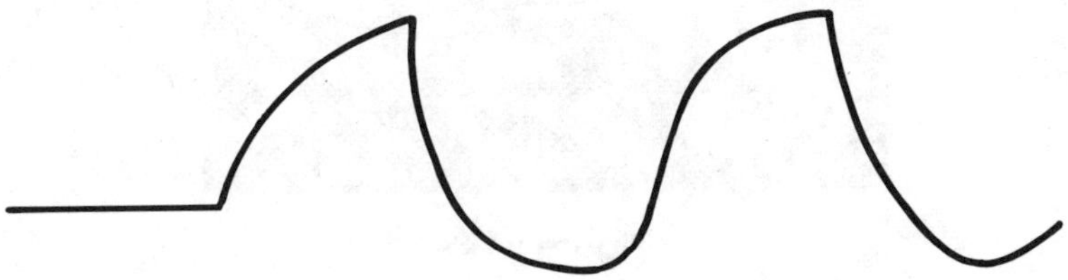

Figure 11-6

The Figures from 11-7 on are actual oscillograms of various responses to the 500Hz square wave. Each is appropriately labeled. These may be useful in evaluating your own system or its components.

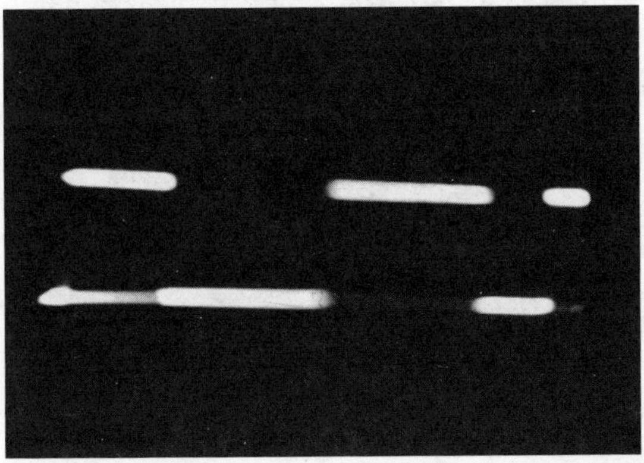

Figure 11-7

Figure 11-7. 3 watt amplifier output across loudspeaker driven by 500Hz square wave generator.

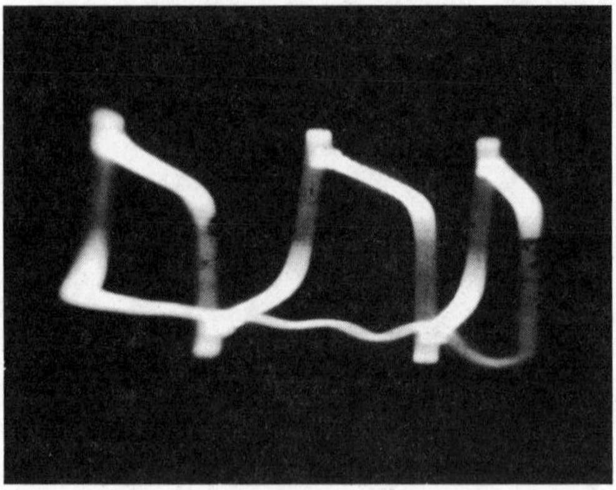

Figure 11-8

Figure 11-8. Playback of tape recorded without equalization showing accentuated high frequency response of the tape.

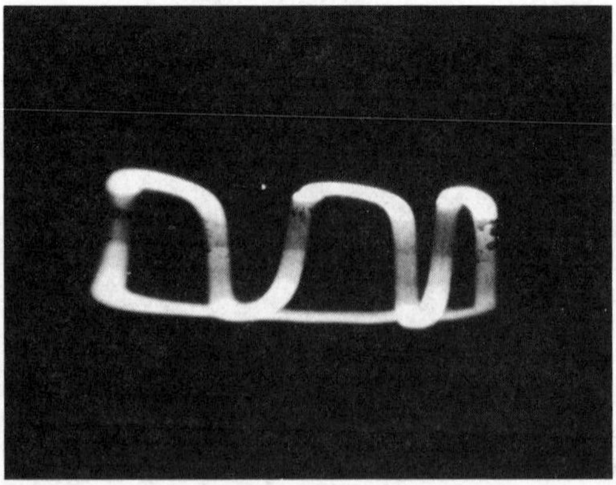

Figure 11-9

Figure 11-9. Playback of tape recorded without equalization showing deficiency of high frequency response of the tape.

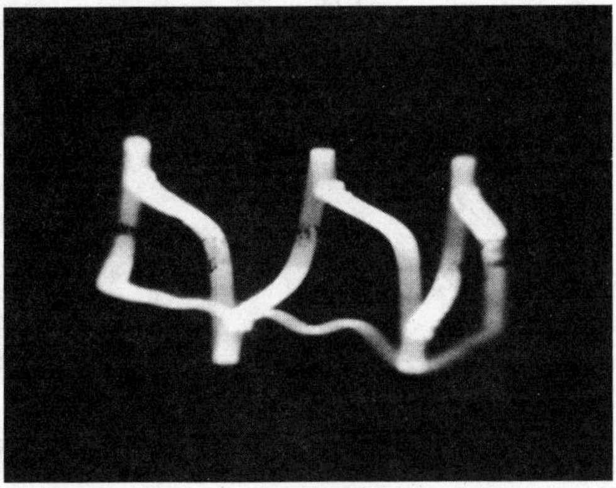

Figure 11-10

Figure 11-10. Playback of a copy of the tape of Figure 11-8 showing build up of high frequency response in tape accentuating high frequencies at 10kHz and above.

Figure 11-11

Figure 11-11. Playback of tape having accentuated high frequency response using equalizer with the 10kHz channel at –6dB.

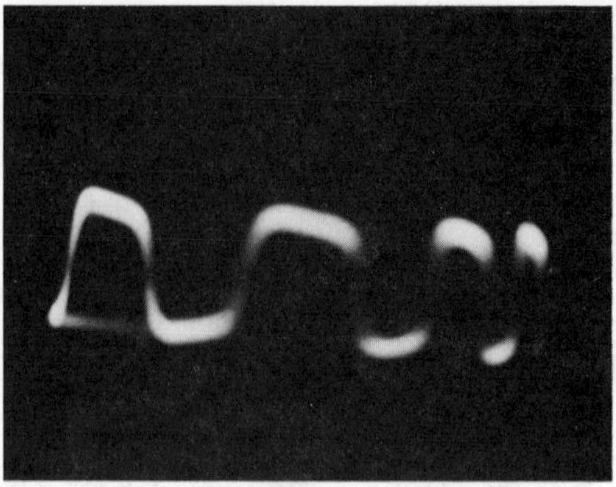

Figure 11-12

Figure 11-12. Playback of unequalized recording made on a commercial prerecorded tape showing inferior high frequency response that is typical of these tapes.

Figure 11-13

Figure 11-13. Steady tone recorded and played back showing irregular response.

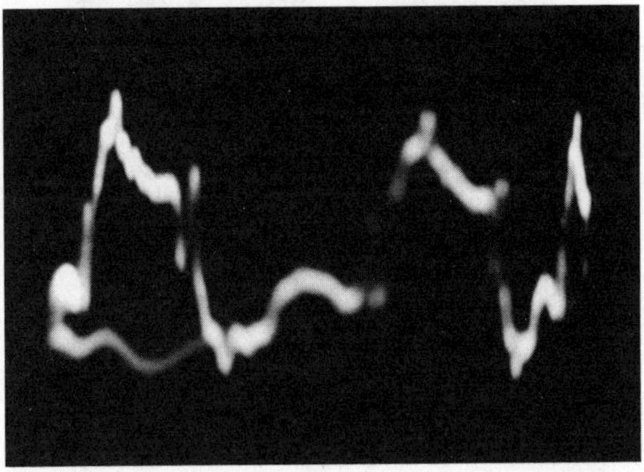

Figure 11-14

Figure 11-14. After you get all through with your fancy amplifiers, tape recorders, etc., this is what you get out of your loudspeaker. Incidentally, this particular loudspeaker sounds pretty good.

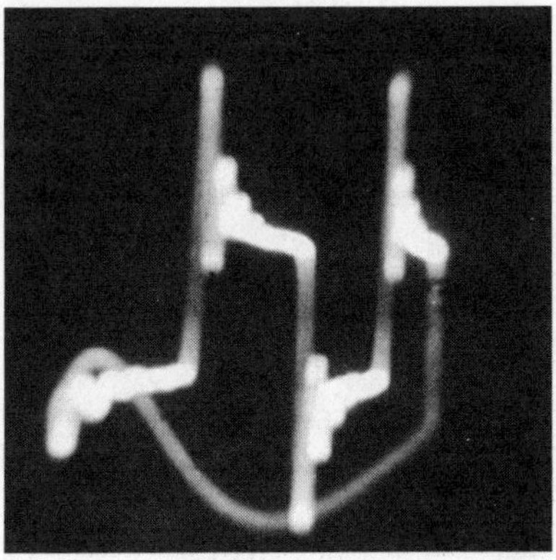

Figure 11-15

Figure 11-15. Square wave phonograph test record played with a pickup which was state-of-the-art 20 years ago.

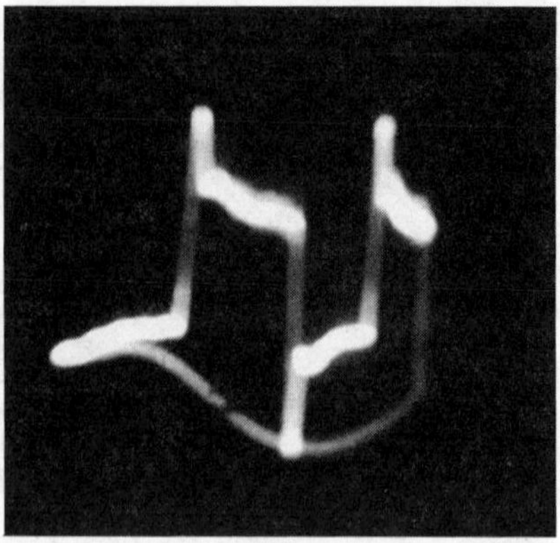

Figure 11-16

Figure 11-16. Square wave phonograph test record played with a current state-of-the-art pickup.

Glossary

ACOUSTICS: Pertaining to the reaction of the environment to audio sounds. Particularly applicable to enclosed spaces such as rooms, concert halls, etc.

AM: Amplitude Modulation. This is the type of modulation in which a radio station radio frequency carrier is varied in amplitude in accordance with audio signals applied to it. In this country AM stations operate in the range from 550 to 1600 kilocycles.

AUDIO: Pertaining to organized sounds to which the human ear responds. Applied to equipment, circuits and devices used in the processing of audible signals.

BAFFLE: A panel or wall imposed between the front and rear of a loudspeaker to prevent cancellation of sounds from the front by sounds from the rear.

BIAS: In magnetic tape recording, the alternating current unmodulated signal that is added to the signal being recorded in order to linearize the magnetic characteristics of the tape.

BREADBOARD: A preliminary and experimental layout and testing of a circuit. Name derives from fact some early experimenters actually used a breadboard for the purpose.

CAPTURE RATIO: The ability of an FM tuner to reject unwanted signals on the same frequency as the desired signal.

CASSETTE: A proprietary tape package patent by Phillips. Self-contained reels of 5/32 inch wide tape adapted for use in suitable machines by merely inserting the package. Generally available in 30, 45, 60, 90 and 120 minute lengths when played at 1⅞ inches per second.

CHASSIS: A metal inverted pan on which electronic circuits and components are mounted.

CIRCUIT: A term applied to interconnected electronic components forming one or more closed loops.

COMPENSATED: Generally means frequency response variations inserted to correct deficiencies in a system or to enhance some desired response.

COMPLEMENTARY SYMMETRY: Generally applied to a transistor power amplifier in which complementary transistors, i.e., an NPN and a PNP, are connected so that one carries positive current and the other, equal negative current to the output load.

DAMPING FACTOR: The ratio of the internal impedance of an amplifier to the impedance of its load. An amplifier with an 0.8 ohm impedance driving an 8 ohm loudspeaker would exhibit a damping factor of 10.

DECK: Usually designates a magnetic tape player/recorder, including the means for driving the tape and playing it back or recording on it. Distinguishes from less sophisticated portable tape machines.

DIP: A plastic integrated circuit package with connecting pins that project along the lower side in two rows. Generally 4 to 16 pins per package. Also sockets for these packages.

DISTORTION: Applied to audio frequency signals that depart from an exact replica of the original signal in wave form or frequency response.

DOLBY: A proprietary system for improving magnetic recording by means of high frequency enhancement on recording and the reverse on playback. May include level controls. Various forms are in use.

DRIVER: A transistor or amplifier for driving a power amplifier. Most power amplifiers require a driver for furnishing power that is not available from a simple voltage amplifier.

EIA: Abbreviation for Electronic Industries Association.

EQUALIZATION: Boosting or lowering the frequency response by means of an electrical network.

FLUTTER: Like wow but at a more rapid rate.

FM: Frequency modulation. The type of radio transmission in which the frequency of the carrier is varied in accordance with the amplitude and frequency of the audio signals to be transmitted. Broadcasting in this country on FM uses carriers from 88 to 108 megacycles.

FREQUENCY RESPONSE: The amplitude of the response of an electrical circuit to signals of various frequencies. In the audio range, the relative response to frequencies from 20Hz to 20kHz.

HARMONIC DISTORTION: When a sine wave signal is changed in form, producing multiples of the frequency of the original signal, such as second, third and so on, harmonics. Caused by nonlinear circuit elements.

HEAT SINK: A device for increasing the heat radiation of a device such as a transistor. Generally, metal fins providing increased cooling by air, by natural convection or by forced blowing.

HIGH FIDELITY: The art and equipment for processing audio signals with a high degree of faithfulness. Above and beyond the average, and meeting or approaching the state-of-the-art.

HIGH-PASS FILTER: A selective electrical circuit which passes high frequency components and attenuates low frequency components of a complex signal.

Hz: Hertz or cycles per second, formerly cps.

INFRASONIC: Frequencies below audibility, about 15Hz. Formerly called subsonic.

INTEGRATED CIRCUIT: Packages that contain two or more transistors along with limited circuit components. Often very complex to the point of equivalency of hundreds of transistors, and in tiny areas on the order of 0.1 inch square. The heart of computers and many sophisticated electronic devices.

INTERMODULATION DISTORTION (IM): The distortion caused by nonlinear circuit elements such that two signals of different frequencies produce additional signals, having frequencies equal to the sum and differences between the original signals.

INVERTING INPUT: An operational amplifier has two inputs, one in-phase with its output and the other out-of-phase. The out-of-phase input is the inverting input. Feedback from the output to the inverting input is degenerative.

kHz: K means thousands. Hz means Hertz or cycles per second. So kHz means thousands of kilocycles per second.

LOW-PASS FILTER: A selective electrical circuit that passes low frequency components and attenuates high frequency components of a complex signal.

MAGNETIC TAPE: A plastic tape—usually polyester or mylar, coated with a thin layer of finely divided magnetic material such as iron oxide or chromium dioxide, and a binder—used for making magnetic recordings.

MAXIMUM POWER: Loudspeaker maximum power is the maximum power recommended that can be safely handled by a given speaker.

MINIMUM POWER: Loudspeaker minimum power rating is generally an indication of relative efficiency of the speaker. The minimum recommended power is that power which will provide a reasonable listening level.

MIXER: An electrical circuit in which two or more signals from different sources are added together in controlled ratios.

NOISE FIGURE (or NOISE FACTOR): The noise that an amplifier or other piece of equipment adds to the desired signal passing through.

NON-INVERTING INPUT: The Input to an operational amplifier that is in-phase with its output, see "Inverting Input."

NPN: A transistor in which a negatively doped semiconductor forms the collector and another the emitter, and the base of positively doped semiconductor is sandwiched between them. Operates with positive bias on collector and base.

OPEN-REEL: A magnetic tape recorder/player in which the tape is carried by open reels as contrasted to the packaged reels of the cassette.

OPERATIONAL AMPLIFIER: A high-gain wide-frequency response amplifier in which the gain is determined by the ratio of a feedback resistor, degeneratively connected from output to input, to the input resistor.

OSCILLATOR: An electronic signal generator in which an amplifier is provided with regenerative selective feedback so that it operates in a self-sustaining anode to generate an AC signal.

OSCILLOSCOPE: A cathode ray tube, as in a television set, including amplifiers and controls for displaying various types of signals on the face of the tube.

PHASE: An AC signal has three primary characteristics: frequency, amplitude and phase. The phase is the relative times at which a signal, for example, crosses zero with respect to the zero crossings of a signal of a different phase.

PHASE DISTORTION: Phase distortion changes the relative phases of various components of a complex signal with respect to their original phases.

PICKUP: In a phonograph or record player, the device carrying the needle and the means for converting needle motions caused by the record groove undulations into electrical signal.

PNP: A transistor in which a positively doped semiconductor forms the collector and another the emitter, and the base of negatively doped semiconductor is sandwiched between them. Operates with negative bias on collector and base.

POWER AMPLIFIERS: An audio amplifier capable of supplying power to drive a loudspeaker. It increases power from some low level to a substantially higher level.

POWER SUPPLY: The device that converts AC from the power line to a direct current of suitable voltage and current capacity to power a given piece of electronic equipment.

PRE-EMPHASIS: A circuit for increasing signals in a given frequency range before they are utilized, as in a magnetic tape recorder.

PROJECT: Complete design and plans for building a piece of electronic equipment, such as an amplifier or a piece of test equipment. This book covers only hi fi projects.

QUIETING: A power reference term used in connection with FM receivers to denote the power input level at which 50dB of quieting takes place. O dBf is defined as 1×10^{-15} watt.

REVERBERATION: A phenomenon in which sound waves in a room, for example, are reflected back and forth to produce a sustaining sound. May also be produced electronically.

RIAA: A standard recording characteristic equalization for phonograph records approved by the Record Industry Association of America.

RUMBLE: Low frequency sounds caused by mechanical vibrations of a phonograph turntable.

SELECTIVITY: The ability of a radio tuner to distinguish between two different stations spaced closely on the dial.

SEPARATION: The ratio of separation of two stereo channels.

SIGNAL PATH: In an electronic circuit, the route that the electrical signal follows in passing from input to output.

SIGNAL-TO-NOISE RATIO: The ratio of the desired signal at a predetermined level to the background noise of a tape, record, amplifier or other piece of equipment.

SOUND LEVEL: The intensity of sound. Most commonly rated in dB in which OdB represents the threshold of hearing and 120dB the threshold of feeling. May be measured with a Sound Level meter.

SQUARE WAVE: As the name implies, a signal having vertical rise and fall and a flat top and bottom. Essentially a transient, it requires a wide frequency response for good reproduction.

STEREO: The sound from two sources that simulates human ears and gives directionality to reproduced sound. Also creates the illusion of space and presence.

THD: Total harmonic distortion. The power ratio of distortion components to a fundamental tone.

TONE ARM: The pivoted arm mounted on the phonograph or record player platform that carries the pickup at its outer end. Its characteristics are very important to the proper playing of disk recordings.

TRANSIENT DISTORTION: The distortion that changes the shape of a transient signal. A transient signal is one that changes shape rapidly as opposed to a steady state signal, which remains substantially constant for a period of time.

TRANSIENT RESPONSE: The response of a circuit or system to a sudden change in signal level.

TRANSISTOR: A tiny solid state amplifying device of germanium or silicon having three terminals, collector, base and emitter. Used individually and in large groups in integrated circuits.

TURNTABLE: In a phonograph or record player, the rotating platter that carries the record.

TWEETER: A speaker reproducing mainly high frequency sounds, such as those above 5-7kHz.

WATTS: A measure of electrical power or acoustic power. Electrical watts are equal to voltage times current times cosine of their phase angle.

WHITE NOISE: Random noise having constant amplitude throughout the audible range.

WIRING DIAGRAM: A drawing showing the physical appearance of components and interconnecting wires of an electronic circuit.

WOOFER: The very low frequency speaker in a sound system that generally provides maximum output in the range of 20 to 200Hz.

WOW: Slow, regular variations in the speed of tape drive or turntable that, if pronounced, cause audible variations in musical pitch.

ZENER DIODE: A diode that, when reverse biassed, suddenly conducts at a predetermined voltage. Used as a reference, regulator or voltage limiting device in power supplies.

Index

A

Acoustics, sound level vs. speaker efficiency, 66
Amplifier
  breadboard three watt, 42
  cost vs. power, 56
  design thirty watt, 57
  how much power needed, 54
  offset, 36-42
  offset considerations, 40-42
  power supply for three watt, 32
  power vs. sound pressure, 55
  practical characteristics, 37
  projects, 31 & 57
  thirty watt, 53-65
  three watt amplifier circuit choice, 36
  three watt outline, 31

B

Book in brief, 5-9
Book's practical value, 5
Breadboard, three watt power supply, 32

C

Cassette magnetic tape, 185
Cathode ray oscilloscope, 141
Circuit diagrams
  reading, 25-27
Circuit diagrams (*cont.*)
  signal path, 29
  typical, 30
Cross-over networks, loudspeaker, 81-87

D

Damping factor, 120-165
  how measured, 163
  measurement set-up, 165
Diagrams
  circuit, 28-30
  wiring, 28
Disk
  low frequency disturbance 96-98
  pre-emphasis, 95
Distortion
  harmonic, 117-120
  intermodulation, 117-121
  meaning of THD, 118
  measuring 118-119
  phase, 121
  significant, 117
  significant levels, 118
  transient intermodulation, 121

E

Electrical tests, 120
Enclosure, loudspeaker, 68
Equalizer
  assembly, 111-112

Equalizer (*cont.*)
  channel responses, 113
  circuit board, 114
  circuit complete, 110
  five channel, build, 108
  front panel, 111
  photos, 115
  theory, 109
  wiring, 114

F

Filter
  loudspeaker, 84-85
  response, 87

G

Glossary of Hi Fi terms, 223-230

H

Harmonic
  distortion, 117-120
  loudspeaker distortion, 69
Hi Fi
  criterea, 17
  limiting factors, 20-22
  projects help achieve, 17
  specifying components, 18-19

I

Intermodulation meter
  assembly and wiring, 127
  basic circuit, 122
  block diagram, 124
  build as a project, 121
  front panel details, 130-131
  calibrating, 133
  circuit details 123-125
  circuit description 123-126
  circuit diagram, 125
Intermodulation meter (*cont.*)
  distortion, 117-121
  theory of operation, 124-126
  using, 134-135
Input meter, build, 155
Input/Output meter
  build, 155
  overload protection, 155
  using, 155, 157

L

Listening room
  bassboost, 208
  criterea, 195
  equalizers, 213
  general, 205-206
  ideal, 206-211
  important, 206
  reverberation, 207
  speaker placement, 208
Loudspeaker
  baffled woofer, 74
  building smaller speakers, 90
  closed box effect, 76
  commercial speaker response, 89
  component selection, 70
  cross-over networks, 81-87
  deadening enclosure, 79
  enclosure significant, 68
  equalization, 168-169
  enclosure construction, 77-80
  filter design, 84-85
  filter response, 87
  free air characteristics, 71-73
  full-size, build, 76
  harmonic distortion, 69
  ideal, 67

Loudspeaker (*cont.*)
infinite baffle, 74
listening room important, 68
measuring phase of current 165-167
measuring power to, 165
midrange, 81
placement for stereo, 210
problems, 67
sound level in baffle, 75
testing, 168-169
three way, 84
three way response, 88
tweeter, 84
two way, 83
venting enclosure, 80
wide frequency coverage, 69
woofer testing, 81

M

Magnetic recording
bias, 182, 196
bias adjustment, 196
care of your deck, 201
cassette, 185
cassette deck, 186
copying speed, 200
deck features, 186
Dolby, 199-200
equalization, 191
evaluating tapes, 194
high frequency response, 182
metal tape, 199
open reel, 185
surface roughness visualized, 202
surface testing with IM meter, 202
tape choices, 187
tapes compared, 193
tape counters, 201
Magnetic recording (*cont.*)
tape speed, 192
tape surface, 190
testing tape, 188-189
Microphone Sensitivity Table, 147
Mixer
preamplifier, 104-105

O

Output Meter
build, 155

P

Phono cartridge
characteristics, 176-177
types, 175
Phono records
good, 177
test records, 178
Power Supply
testing three watt, 34
thirty watt amplifier, 59-61
three watt amplifier, 32
Preamplifier
circuit board wiring, 101
circuit diagram, 98
disk pre-emphasis, 93
mike, auxiliary and phono, 98
mixer provision, 104-105
mixer wiring, 106
requirements, 93
tape and disk compared, 94
tape pre-emphasis, 96-97
versatile, 103
Project building
helps achieve great Hi Fi, 17
materials, 24
tools, 23

Projects to build
amplifier, thirty watt, 53-65
amplifier, three watt, 31-36
equalizer, five channel, 108-115
input/output meter, 155
loud speaker, full size, 68-85
output meter, 155
power supply, for 30 watt amplifier, 59-61
power supply, for 3 watt amplifier, 32
preamplifier, mike, phono & aux., 93-105
sound level meter, 146-149
square wave generator, 137-138 & 216-222
three tone test oscillator, 151-155
white noise generator, 143-144

R

Reverberation time
defined, 210
Sabine's book, 209

S

Sound level meter
adjusting, 148
breadboarding, 145
building, 147-148
calculating gain, 147
choosing the design, 145
circuit diagram, 146
wiring diagram, 149
Speakers (*see* Loudspeakers)
Square Wave
affected by frequency response, 215
construction, 138
Square wave (*cont.*)
illustrated, 216-217
oscillogram, 217-222
reveals response characteristics, 137
testing revealing, 137
three watt amplifier response, 47
Stylus, types, 176
Symbols
circuit diagram, 25-27

T

Tape decks (*see also* Magnetic recording)
magnetic recording notes, 181
Testing projects
general, 159
loudspeakers, 168-169
thirty watt amplifier, 167
three watt amplifier 161-162
three watt power supply, 34-35
Thirty watt amplifier
alternate circuits, 58-59
build, 61-62
circuit board layout, 62
design, 57
enclosing, 63
photos, 64-65
power supply, 59-61
power supply determining, 53
Three tone test oscillator
applying, 155
building, 152
choosing the circuit, 151
circuit chosen, 152
final circuit, 151
front panel, 153

Three tone test oscillator (*cont.*)
wiring diagram, 153
Three watt amplifier
chassis, 47
enclosing, 45
frequency response, 46
front panel layout, 47
interconnections, 48
intermodulation distortion, 50
layout, physical, 43
outlining, 31
power supply layout, 33
square wave response, 47
testing breadboard, 44
wiring diagram, 44
Tone arm
antiskating, 174
characteristics, 173-174
tracking error, 174
Tools for project building, 23
Total Harmonic distortion (THD), 120
Transient intermodulation distortion, 121
Troubleshooting projects
general, 159
step-by-step, 160
tracing signal, 160-161
Turntables
characteristics, 172-173
specifications, 172-173
Tweeter, loudspeaker, 84

W

White noise generator
building, 143
circuit and wiring, 144
components, 143
white noise defined, 143
Wiring
diagrams, 28
Woofer, loudspeaker, 74
Words from the author, 5-9